PREDICTIONS

PREDICTIONS

Edited by Sian Griffiths

With an Introduction by Jonathan Weiner

OXFORD
UNIVERSITY PRESS

Great Clarendon Street, Oxford OX2 6DP

Oxford University Press is a department of the University of Oxford.
It furthers the University's objective of excellence in research, scholarship,
and education by publishing worldwide in

Oxford New York

Athens Auckland Bangkok Bogotá Buenos Aires Calcutta
Cape Town Chennai Dar es Salaam Delhi Florence Hong Kong Istanbul
Karachi Kuala Lumpur Madrid Melbourne Mexico City Mumbai
Nairobi Paris São Paulo Singapore Taipei Tokyo Toronto Warsaw

with associated companies in Berlin Ibadan

Oxford is a registered trade mark of Oxford University Press
in the UK and in certain other countries

Published in the United States
by Oxford University Press Inc., New York

Introduction © Jonathan Weiner 1999
For further copyright details see pages 321–2

The moral rights of the author have been asserted

Database right Oxford University Press (maker)

First published 1999

British Library Cataloguing in Publication Data
Data available

Library of Congress Cataloging in Publication Data
Data available

ISBN 0–19–286210–3

3 5 7 9 10 8 6 4 2

Typeset by RefineCatch Limited, Bungay, Suffolk
Printed in Great Britain by
TJ International Ltd, Padstow, Cornwall

Contents

Preface

'**N**ever fall in love with your own airship', warns Umberto Eco on page 105. In other words, weave theories and build inventions—but don't take your creations too seriously. For who can foresee what will survive the passage of time?

Fortunately, the 30 thinkers featured here have largely set aside their reservations about foretelling the future. First profiled in a series for *The Times Higher Education Supplement*, for this book all 30 were asked to supplement those original interviews with a prediction for the 21st century. Each was asked what scholarly breakthrough they would most welcome before the year 2100, and how it might impact on society.

Their predictions encompass fields as diverse as biology and politics, philosophy and physics. The possibilities they sketch for the next century are manifold: children genetically engineered to resist new and deadly viruses, the creation of 'bionic' people capable of understanding thoughts without the necessity of language and a solution, finally, to the riddle of human consciousness. We will create life in our labs and send it out to thrive on hostile planets, elect female governments to put an end to war and teach schoolchildren how to manage their emotions just as today we teach them English and Maths. Reason and democracy will triumph and animals 'will be legally recognized as sentient beings, with rights that can be enforced in the courts by guardians acting on their behalf'.

But over such brave new visions shadows loom. How will we achieve such changes and what risks might they bring? Is mankind really no more than 'a kind of lethal mutation', as Noam Chomsky suggests? Will we even survive the next 100 years?

Without the predictions of the 30 thinkers profiled and the work of the writers who interviewed them this book would not have been possible. Thanks to all of them for their writing, time and patience.

Nor would it have happened without the efforts of Sonya Allen, Michelle Blore, Ed Fitzgerald and Tricia Grogan at The Times Supplements, and senior commissioning editor Angus Phillips and assistant editor Helen Cox at Oxford University Press.

Thanks, too, to the photographers who allowed their pictures to be reproduced and to The Times Supplements picture editor Jane Harper and designer Kathryn Coates.

And a special thanks to Adam Harper, Martin Ince, Sarah Knowles, Val Pearce, Andrew Robinson, Tom Wakeford, Lynne Williams, and Katrina Wishart.

And finally, thanks to THES editor Auriol Stevens, who, despite her own reservations about futurology, supported the book from the very beginning.

Sian Griffiths

Introduction

JONATHAN WEINER

Tibetan diviners emptied their minds and stared into lakes, mirrors, the flaming wicks of lamps, or the balls of their own thumbs. Chinese diviners inscribed questions on tortoise-shells, then roasted the shells and interpreted the cracks. Arabian geomancers drew dots in the sand and assigned each set of dots to one of fourteen geomantic Houses, including The House of the Questioner, The House of the Result, and The House of the Result of the Result. To randomize the dots, geomancers sometimes whirled around and around and jabbed their fingers into the sand as they fell—gyromancy.

Meanwhile, Roman diviners who read the entrails of sacrifices were known as haruspexes, literally Gut Inspectors. Cicero wondered how one haruspex could ever look into the eyes of another without laughing.

Today, all these ancient methods of predicting the future look like confessions of impossibility. Several centuries of experimental science have taught us to distinguish between the predictable and the unpredictable. We have learned that there are questions we can answer, and others that can be answered only by the events themselves. Drop a bottle of champagne off the edge of a table, and you know where it will land and what will happen to it. But drop a bottle off a ship's railing, and while you may make an educated guess about the currents and the winds, you can't predict where the bottle will wash ashore or whose hands will reach down at last to pick it up.

This year, leaning on the railings, approaching the most celebrated invisible line that any of us will sail across in our lifetimes, we know that we can't know all that we would like to know about the shape of things to come. We try to look ahead, because we are still human, but many of the questions we would like to ask in 1999 are unanswerable in 1999. The fate of most centennial predictions is the fate of most messages-in-a-bottle: they just get lost.

For us, schooled and chastened by the scientific process, the best way to think about the future is not in terms of the magic sudden knowledge of the diviner but the slow hard-won knowledge of the experiment. The best way to think about the future is to try to pick out a few of the future's greatest experiments—experiments that are already in progress and that are too big for anyone to stop. If they are true experiments, then by definition their results cannot be predicted in advance; still, if they are good experiments, then the answers will be interesting. We might try to talk about the next century as a set of five experiments in progress, experiments that will certainly have results, results that will assuredly be interesting.

The first and most fundamental experiment in both our century and the next is probably the human population explosion. Our numbers reached 6,000,000,000 at the close of the twentieth century, having nearly quadrupled from the beginning of the century to the end, with 1,000,000,000 of us now living in India alone. 6,000,000,000 human beings is a global experiment.

A second experiment, coupled to the first, is global warming. Several decades have now passed since an earth scientist first thought to call this a global experiment. Each year as the planet's temperature inches upward the experiment is getting more interesting and absorbing to watch.

A third experiment, entwined with both the first and second, is the progress of technology, which promises and threatens so much. We are married to this experiment and we often have cause to remember the phrase 'for better and for worse'.

A fourth experiment, again involving all the others, is the progress of evolution. Our species now has the power to shape the evolution of many others around us, and perhaps our own evolutionary future too.

A fifth experiment is the one that the evolutionist E. O. Wilson has recently called consilience. This is the dream, hope, or prayer that everything we are learning about ourselves and our universe will fit together someday soon in a single overarching vision, comprehending all of our ancient, modern and post-modern philosophies, all our arts and sciences, all our experiments, and all our inmost feelings, including the mixed feelings of hope, dread, curiosity, practicality, and pure play that drive us to try to look ahead.

Predictions brings together the thoughts of thirty of the world's most distinguished minds. They have devoted their careers to these global experiments, and here they make their highly educated guesses about the results. Most of them take pains to acknowledge that they cannot foretell the future by gazing at the balls of their thumbs. The philosopher Daniel Dennett quotes H. L. Mencken's observation that 'for every difficult question there is a simple answer—and it's wrong'. The chemist Carl Djerassi suggests that since the answers will not be black and white, the twenty-first century should adopt grey as the colour of its banner. This is a winsome suggestion, like Benjamin Franklin's proposal that the turkey should be the national bird of the United States, and equally unlikely to be taken up. The Italian novelist and philosopher Umberto Eco recognizes that those who are most drawn to look ahead are also most in danger of failing. Many aviators once dreamed that the best ships to fly through the air were lighter than air, until the disaster of the Hindenberg. Today virtually everyone agrees that the best flying machines are heavier than air, not lighter. Eco writes: 'The moral of the story is that in both philosophy and the sciences you must be very careful not to fall in love with your own airship'.

More than one of the academics in this volume has already helped to shape our global experiments. Djerassi, the chemist, may have altered the course of the population explosion, at least slightly, by inventing the oral contraceptive. The Pill changed the social climate of the 1960s and helped to ensure that what is known in the U.S. as the baby boom was not followed by a second boom a generation later—only a baby boomlet. On the other hand, as Djerassi himself points out, Italy and Spain have low fertility rates today, and neither country is a major consumer of his pill. So in the population experiment, social and cultural chemistry seems to matter more than laboratory chemistry.

The future of the baby boom is a political issue in the U.S. Baby boomers wonder if their country will be able to support them when they retire. The linguist and political activist Noam Chomsky argues that this is just a scare concocted by Wall Street to distract gullible baby boomers. 'It takes five minutes' thought to see that that problem was dealt with when they were children,' he says. If the U.S. could afford to take care of them as babies then the country will surely be able to take care of them again when they are growing old. 'This isn't quantum physics,' Chomsky says. 'Any school student can figure this out'. Being a baby boomer myself, I was encouraged by Chomsky's argument, and I repeated it over dinner the other night to our two schoolboys. Our thirteen-year-old wondered if we can equate the cost of raising children and the cost of caring for the old. It takes a village to raise a child, but it takes a village, three doctors, five consulting specialists, and six hospital beds to care for an old man.

Is Chomsky right or wrong? An economist could provide some figures to help us here. But the future is a lot to see in a five minutes' tour. There is the House of the Result, and the House of the Result of the Result.

Most population experts now believe that the growth in human numbers will level off in the mid-twenty-first century, for the same poorly-understood reasons that growth is levelling off now in so many technologically advanced countries. Among Western social theorists, fears about the population explosion seem to have been supplanted lately by hopes for the growth of democracy. The Cold War that polarized the planet in the decades after World War Two has given way to globalization, a trend so powerful that, as the analyst of public policy Francis Fukuyama says, 'It is very hard to see how that is going to be undone'. Even countries on opposite sides of the globe are practically neighbours now, and people everywhere are growing more and more alike in their ways of looking at the world. Fukuyama repeats his famous prediction that the flourishing of liberal democracies means the end of history. The economist Amartya Sen agrees that democracy has now come to seem the normal state of government. Where the nineteenth century looked at a poor country and asked if it was ready for democracy, the late twentieth century expects virtually every country to be capable of 'becoming fit, as it were, through democracy'.

Most of the social thinkers in this volume sound optimistic about the next century. Fukuyama believes that feminism along with globalization will help keep a crowded planet out of war. Andrea Dworkin, who was radicalized in her twenties by being beaten by her husband, is cautiously optimistic about the future of women in the twenty-first century. She does not expect the century to bring the death of patriarchy, but she does predict more and more resistance from women around the world. Even the Nigerian novelist Chinua Achebe, author of *Things Fall Apart*, considers himself an optimist today, for reasons that are somewhat similar to Dworkin's. He says his country has seen such horrors, 'the excesses of bad government which lie like a curse on the continent', that he now believes these nightmares will serve as correctives for the future. 'We're not good students, but in the end we do pick up pieces here and there. This is the hope, the only hope, perhaps'.

The atmospheric chemist F. Sherwood Rowland deserves the whole world's gratitude for helping to save the ozone layer and in the process, slowing the progress of global warming. In the early 1970s, Rowland and a colleague realized that a single chlorine atom released into the stratosphere can destroy as many as 100,000 ozone atoms. This multiplier effect meant that aerosol cans using chlorofluorocarbons as propellants could seriously damage the ozone layer. When a huge ozone hole was discovered above the south pole in the mid-1980s, the discovery—and Rowland's campaigning—helped lead to a world ban on the use of chlorofluorocarbons, which also happen to be greenhouse gases. Rowland predicts that some time in the next century, when the load of carbon dioxide begins to produce global changes as shocking as the ozone hole, we will begin at last to sequester our carbon dioxide, muzzling smokestacks and exhaust pipes and burying the carbon—that is, trying to put it back where we found it.

As Rowland points out, we now get 85 percent of our industrial energy by burning fossil fuels. Back in the days when labour was limited to what each human being could do with two hands, and maybe two oxen, a single worker could do very little to change the planet. With fossil-fuel-driven machinery, each worker can do much more work. So the global experiment in the air is driven both by our numbers and by the increasing power of our technology—another multiplier effect, and the outcome of the first and second experiments depends in part on the outcome of the third. One of the best places to try to glimpse that is the Massachusetts Institute of Technology, where a team of computer engineers and consultants (including Dennett, the philosopher) is building one of the world's most sophisticated robots: Cog, six feet tall and designed to learn like a baby. If we are lucky, electronics itself will turn out to be in its infancy, and our engineers' explosive experiment in miniaturization will help someday to reduce technology's impact on the planet.

That is, if we are very lucky. Before it saves the global ecosystem, electronics may swamp the human nervous system. The computer scientist Kenneth Warwick and the designer Don Norman foresee a future in which ergonomics will be extended and turned inside out. Call this organomics. Instead of designing machines to fit the human body, the next generation will design machines to fit inside the body, redefining human anatomy and what it can do. In 1998, Warwick had a silicon chip implanted in his arm so that as he approached doors they would open and as he approached buildings they would say, 'Hello, Kevin'. He hopes to link his wife's brain and his own by computer chips so that they can read each other's thoughts. Likewise, Norman dreams of cellular telephones and simultaneous translators so small that they can be surgically embedded beneath the flap of the ear, along with web browsers, calculators, modems, and memory chips to enhance our brains.

I myself wouldn't even wear a wristwatch if my wife hadn't given me one for Father's Day. As I thought about the organomic vision of the future, just now, my brain remembered and sang to itself (without any help from chips) the refrain of a Loudon Wainright song: 'I'd rather be lonely'.

Our fourth experiment, which has been called volitional evolution, will be accelerated by the mapping of the human genome. That map—it seems safe to predict—will be one of the great scientific achievements that inaugurate the new century. It is already leading to new dreams for medicine. French Anderson, director of gene therapy at the University of Southern California, is leading an effort to treat ADA deficiency in the womb. He predicts that gene therapy will have revolutionized the practice of medicine by the year 2030: by then there will be a gene for every disease that flesh is heir to. Anderson believes that eventually we should begin adding and rewriting genes in sperm and eggs as well, making changes not just for one child, as he acknowledges, 'but for grandchildren, great-grandchildren, great-great-grandchildren and so on', which would take us well beyond the twenty-first century—messages in bottles that will drift across the next millennium. He is probably right that we will in fact do this. We will do it as soon as soon as doctors feel that they can inject genes into sperm and eggs while honouring their professional code, 'First of all, do no harm'. 'When the time comes, then we must do it,' Anderson says, 'because it is just plain human nature'. Of course, we'll be changing human nature too, bit by bit.

For our self-absorbed species, this may become the most alarming experiment in progress: the conscious and deliberate revision of human evolution. How well it goes will determine how the people who brought it to us in the twentieth century will be remembered. No other experiment pits so many of the people in this book against each other. Anderson notes that Stephen Jay Gould, 'whom I have known for years', is on the board of an American organization, the Council for Responsible Genetics, which put out a broadside against Anderson's ADA proposal under the headline 'Say No to Designer Children!!!'. The designer Don Norman wonders if planning for a baby will someday be like ordering a new car: 'the proud parent(s) (singular? plural?) will be able to choose from a list of options', everything from hair colour to brains. James Watson retorts, 'People say we are playing God. My answer is: "If we don't play God, who will?"'.

Watson himself is more worried about cloning human beings than about rewriting human genes. But I don't understand why, since cloning can only produce copies of an experiment that has already been tried, whereas rewriting can produce a new genetic experiment.

We are also rewriting the genes of our crops and our very pests and plagues. Although the economist John Kenneth Galbraith warns of the dangers of the nuclear stockpiles, the worst terrorist blasts in the next century may come from a terrorist's test tube. This is a threat before which we all stand naked. Our only hope of refuge must come from a growing humanity and a future in which, as Daniel Goleman writes, 'empathy will hold as valued a place in the curriculum as algebra'.

Many of the academics in *Predictions* hope that the study of the mind in the next few decades will help to unite past and present, biology and culture, nature and society, matter and consciousness, and the natural and the human sciences. 'A fundamental division between the humanities and sciences may become as obsolete as the division between the celestial and terrestrial spheres,' writes the cognitive scientist Steven Pinker.

This fifth experiment, the experiment of consilience, is in a sense an optimistic assessment that all the other experiments on the planet will add up. Certainly it is a reasonable prediction that the biotechnologic and electronic explosions we are witnessing today will be followed by explosions in the humanities. We may hope that these revolutions will fit together but it is also possible that they will continue flying apart. The evolutionist Richard Dawkins has compared the origin and spread of ideas to the origin and spread of life. One depends on the self-replicating bits of information we call genes, and the other depends on the self-replicating bits of information he calls memes. Ideas are now multiplying and spreading from human brain to brain the way genes multiplied in the early earth from sea to sea. 'You have, in effect, a new primeval soup,' says Dawkins. Dennett takes up Dawkins' theme in *Predictions*. Just as European germs once blighted new continents wherever they spread, it is possible that during the next century, it will be 'our memes, both tonic and toxic, that will wreak havoc on the unprepared world'. If the golden arches and kudzu of capitalism go on spreading, Dennett predicts more and more global cultural disasters. 'The field of "public health" expanded to include cultural health will be the greatest challenge of the next century.'

Fukuyama marches in the other direction. He hopes that genetic engineering 'may make possible certain kinds of social engineering that heretofore have failed'. His hobby, he confides, is building ever fancier virtual furniture on his personal computer. We should remember that when a successful futurist plays with virtual reality late at night, the next generation may be sitting on his furniture.

If medicine succeeds in prolonging life as promised, some of us may still be here to judge the success of the global knowledge experiment at the close of the twenty-first century. From here in 1999, its future is hard to guess. In the next-to-last chapter of *Predictions*, the cosmologist Steven Weinberg dreams of a final theory, a 'Theory of Everything', and in the last chapter the philosopher Slavoj Žižek dreams that we will never understand our unconscious: 'That unknowability spells freedom'. *Predictions* also includes the thoughts of one of the world's most famous living futurists, Arthur C. Clarke, who foresaw the invention of communications satellites in 1945 and in his story 2001 imagined one of his heroes on his way to the Moon reading electronic newspapers on a 'foolscap-sized newspad' that sounds very much like a Powerbook. But even Clarke himself would rather be known as an extrapolator than a prophet. 'Often,' he confesses, 'it has been more interesting to see where (and why) I went wrong, than where I happened to be right'. He doubts that the ultimate questions will be answered in a century or even in a millennium. 'It is not likely that ultimate questions will be settled in such a short time, or that we will really know much about the universe while we are crawling around in the playpen of the solar system'.

Clearly all of the predictions in *Predictions* can't be right, because so many of the contributors disagree with each other. When Dawkins was asked by students at Cambridge University to speak alongside his fellow-evolutionist Lynn Margulis, he snapped, 'I'd rather share a platform with Attila the Hun!'. They do share a platform here. Daniel Goleman, author *of Emotional Intelligence,* praises the power of restraint and delayed gratification, of postponing the marshmallow, while Dale Spender celebrates the virtual world's instant and anarchic gratifications, and she believes the future is on the computer's side: 'Students will vote with their virtual feet'. As a boy, Dennett, the philosopher, rigged up an alarm system for keeping his sister off his territory. As an adult he says he has often fought with those in neighbouring territories. 'I have tried to make it a rule that I don't slam people unless they are really big and they really deserve it,' he says. In the past, Dennett has slammed Gould and Chomsky. Gould has slammed Dennett and Dawkins. Alarm bells will be ringing for years inside this little time capsule, adrift on the currents it attempts to predict.

How will it all add up? The greatest dream of the palace of reason, the dream of the European enlightenment thinkers, is dreamed almost everywhere now. In a sense, as Amartya Sen writes, it always has been dreamed, the concept of enlightenment having been invoked in ancient Sanskrit, Pali, Chinese, and Arabic, and the very name of Gautama Buddha meaning 'The Enlightened One'. So the outcomes of the five global experiments will be watched by more and more people in the next one hundred years. But whether the next century will be an improvement on the last, or whether any of the specific predictions in *Predictions* will come true is impossible to say. Guessing the fates of any single one of the predictions in this book is like guessing the path of the bottle in the ocean, or the fate of any one of the millions of species that grow and jostle on the tree of life. As Darwin wrote about that problem, 'To attempt to follow the mutual action & reaction in any one case, would be as hopeless as to throw up a handful of feathers on a gusty day & attempt to predict where each particle would fall'.

Predictions

Chinua Achebe

HINUA ACHEBE DREAMS OF HOME EVERY DAY. 'I CAN'T IMAGINE A SITUATION IN WHICH NIGERIA WOULD CEASE TO FEEL LIKE HOME,' HE SAYS. 'OUR EXILES ARE DIFFERENT FROM OTHER EXILES IN THIS RESPECT. THERE IS ALWAYS THIS CONNECTION TO THE BASE. WHEN NIGERIANS GET TOGETHER, THEY REALLY ARE TALKING ABOUT NOTHING ELSE. EVEN WHEN THEY ARE NOT TOGETHER, YOU KNOW, IN LITTLE FAMILIES, THIS IS WHAT THEY ARE VERY MUCH OCCUPIED WITH.'

Achebe's 'little family'—himself and his wife Christie—spend their days in a wooden clapboard bungalow on the campus of Bard College, 100 miles north of New York City. The walls of the house are covered with Achebe's many awards and honorary doctorates—including Nigeria's National Merit Award—and his study displays a large map of the continent of Africa, but despite these efforts to make himself feel at home, his working environment, set back among the trees and the chipmunks on the shores of the Hudson River, seems very American.

Achebe's present exile began in 1990 when he suffered a terrible motor accident. The Nigerian hospitals were unable to cope with his injuries, so he was flown to a hospital in Britain. After six months he moved to America, partly to a professorship in literature at Bard College but mainly—since he still has to cope with semi-paralysis—in order to be close to excellent medical care in the States. He expected only to stay for a short time, while he convalesced, but then the political situation began to deteriorate in Nigeria.

The country's leader since 1985, Ibrahim Babangida, kept postponing a return to civilian rule but finally allowed a democratic election in 1993. Moshood Abiola was elected by popular vote but the election was annulled and a weak interim government was set up. It was overturned in a military coup led by General Sani Abacha and Abiola was imprisoned. Abacha—described by fellow Nigerian Wole Soyinka as 'a gloating sadist and self-avowed killer'—then instigated the harshest regime that Nigeria has ever known, arresting and arbitrarily executing many leading figures, including the writer Ken Saro-Wiwa. Most of the country's leading intellectuals and professionals—the writers, the teachers, the doctors—have been driven to leave the country and Nigeria's key institutions were left barely functioning. 'I don't like to call myself an exile in the sense it means for many people,' Achebe says. 'But at the same time, Nigeria would be very difficult for me to live in now with the hospitals broken down and so on.'

Exile is particularly difficult for Achebe, who is still emeritus professor of literature at the University of Nigeria, because his writing is so rooted in the African community. Indeed he was driven to write his first and most famous novel, *Things Fall Apart*, published in 1958, because he felt that nobody was really describing the Africa he knew. 'Over a period of time I was aware that my real story had not been told. I had read novels but nothing I saw was about me. Especially novels which were written about the African people. I didn't recognize immediately that these were supposed to be people like myself.' At first, Achebe says, this lack of fiction about Africa did not seem odd, since all his education was based upon the British curriculum and about subjects far removed from his experience. 'You wouldn't believe, for instance, that the geography lessons I had were about the Vale of Evesham,' he laughs.

Things Fall Apart is not about a harmonious and happy African community but about a community breaking down and under threat. It focuses upon the fate of Okonkwo, one of the leaders of the Igbo tribe who possesses all the virtues of strength and courage which were traditionally admired by his people but who cannot adjust to the changing customs of his time—changes brought about by the arrival of the first white man in Africa. Christian missionaries cause division in the society, as some Igbo people, like Okonkwo's own son, convert to the new religion while others, understandably, resist. Okonkwo feels that he is carrying the community with him, as he always has done in the past, when he kills a white man. But he quickly discovers that his tribe has moved on and that, in his mad act of killing, he has isolated himself from his community forever.

Unable to live with the shame, he commits suicide. His hanging body is cut down from the tree by white men as his own people cannot touch it. 'It is an abomination for a man to take his own life,' one of the characters explains to the white District Commissioner. 'His body is evil and only strangers may touch it. That is why we ask your people to bring him down, because you are strangers.'

Achebe's depiction of Okonkwo, caught between two worlds, was probably inspired by his childhood experience, growing up in a very Christian household but surrounded by more traditional neighbours. He too encountered a clash of cultures, although he now recalls this clash as leading more to good humour and amusement than to the violent conflicts of *Things Fall Apart*. His parents were both pioneer Christians and tried to keep a strict control over their children's contact with those they considered heathen. But Achebe became fascinated by the traditional stories and rituals of his people. 'I was perfectly certain that I was going to heaven, but the masquerades, the stories—I liked those as well. They were not offered as part of my education. So I found myself sneaking out to find out what the story-tellers were doing and I discovered that they weren't people very different from me,' he remembers.

The tensions and divisions in Achebe's childhood world were evident particularly when it came to food and eating practices and, just as now in America, he found himself crossing between the two cultures. 'If my parents knew that we had gone to one of the neighbours on a feast day and eaten there, they would not have been happy with us,' he reflects, 'because we were told that these heathen offered their food to idols. I never saw anyone offering food to idols. Probably what they meant was dropping their palm wine on the ground for the ancestors. To Christians of my father's generation, that's just an abomination. So they said, "Don't eat in the houses of non-Christians because we don't know what they do with their food." But I took my younger sister across when my parents weren't looking and we ate and there was nothing wrong with their food.'

The story of Okonkwo became an instant success. At last African people felt they had their own account of their lives. Published just before Nigerian independence at a time when various local newspapers were springing up, the novel became crucial to the discovery of African self-identity. Elizabeth Ohene, brought up in Ghana and now a journalist for the BBC World Service for Africa, remembers the impact the book made upon her as a child. 'It was the first novel I'd read which captured the atmosphere of life as I knew it,' she says. 'It was supposed to be Literature (it was a school set book in Ghana) and yet I could relate to it immediately. I knew some of the characters that he was describing and I knew what he was writing about.' Even now, according to Ato Quayson, director of the African Studies Centre at Cambridge University, *Things Fall Apart* is 'the most widely read and cited book in African literature.'

The book also brought African literature to international attention. Published by Heinemann, it inaugurated a new series of African writing for which Achebe went on to become the General Editor, playing a seminal role in discovering and encouraging other new African authors. The novel is read across the globe, particularly in countries which are confronting their colonial past. Schoolchildren in Korea, for example, have recently written to Achebe, telling him of the novel's significance for them when they think of their colonization by Japan. 'Stories have a way to go where they want to go,' he says.

After publishing *Things Fall Apart*, Achebe went on to publish two more novels in quick succession in the 1960s, *No Longer at Ease* and

Arrow of God, which traced the history of the Igbo people's relationship with the White Man after the time of Okonkwo. His next novel, *A Man of the People*, had a much more recent setting, depicting a newly independent, imaginary African country in the 1960s. African politics, represented in the novel by the fictional leader Nanga, are corrupt and chaotic; order is only established at the end through a military coup. Achebe's novel was published in January 1966, a few days before the attempted military coup of the idealistic Major Nzeogwu in Nigeria.

Following the second military coup of that year, which established the harsh regime of Yakubu Gowon, dedicated to crushing—and killing—Eastern Nigerians, Achebe found himself personally hunted by soldiers. Because of the ending of *A Man of the People*, he was thought to have been closely connected with Nzeogwu. Forced into hiding, he later smuggled his way out of the then Nigerian capital Lagos to his home region of Eastern Nigeria.

When, in 1967, Eastern Nigeria claimed its independence as the Republic of Biafra, Achebe played a key part in the politics of the new regime, working at the Ministry of Information and, later, as chairman of the National Guidance Committee, speaking and writing about the fate of Biafra. The war between Biafra and Nigeria touched him directly. Nigerian planes bombed his house and his best friend, Christopher Okigbo, who had helped him set up a publishing house called Citadel Press in Biafra, was killed. 'The civil war was a very traumatic experience for me,' he says. 'It was violent and bloody and we were absolutely stunned.'

After the defeat of Biafra in 1970 Achebe returned to his post at the Institute of African Studies in Nsukka, but he found that those who had been involved in the Biafran struggle were being covertly punished. His request for a passport was repeatedly refused and his freedom curtailed and so when the invitation came in 1972 from the University of Massachusetts to take up a visiting professorship, he decided to leave. He spent four years in the USA, during which time he says he felt too sad and disorganized to write. Only once General Gowon's regime was overturned by another military coup did he think about returning and he finally accepted a chair at the University of Nigeria in 1976.

Now he has become one of the key voices of the continent. At the London literary festival last March, he appeared alongside Wole Soyinka and Derek Walcott at a big jamboree in the Hackney Empire billed as 'Two Nobel Laureates and a Legend' and was introduced during the evening by a representative from the British Council as the writer who has become a 'legend in his own lifetime' and who should have won the Nobel prize. Always quiet and modest, Achebe prefers to stress the responsibility of a writer to his community.

Imaginative writing is crucial to the identity and prosperity of any people, Achebe believes. 'It's very important that we understand how our ancestors dreamed the world,' he says. 'Even when the practices have gone, there are still remnants of the thought behind it which are active in our lives and we do not realize it. Finding stories is not only to show that you were kings in the past. You might find things that will explain why you haven't done as well as you should have and that's also as valuable.'

Achebe's role as the voice of Africa means that he is called frequently to comment not just on the literary events of the country but also on the turbulent social and political situation. Last year, for example, he gave a lecture to the World Bank in which he called for the cancellation of Africa's debt. And he writes essays and articles both in English and in Igbo for his two audiences.

He is also working on a novel, his first since *Anthills of the Savannah*—itself published in 1987, over twenty years after his previous novel. Just as during his last period of exile, he is finding it difficult to write the novel when he is living so far away from his native land and the chief resources of his imagination. So is he thinking of returning, especially now in the new democratic era of Olusegun Obasanjo's government, sworn in on 29 May 1999? 'Many people want me to come back,' Achebe says. 'They understand why I am not there. But there is almost a plea to come back and this is something I cannot really ignore.' Hedging his bets, but perhaps very soon he will return.

by Jennifer Wallace

The Colour-Line

CHINUA ACHEBE

I am optimistic about the prospects for Africa in the twenty-first century. I always think that it is better to be optimistic than pessimistic because, if you are hopeful, the worst that will happen is that there might be some disappointment. But if you are pessimistic, you would be miserable all the way.

I don't suggest, however, that simply being optimistic is ever enough. There is work to be done. It is this work that will create the possibility of change for the better. If one does not do this work, then no change will come about. This is more or less the idea at the end of my novel *Anthills of the Savannah*, where you have a situation similar to that which you have in Nigeria now. So many people have been killed or removed from the scene. But there seems to be an opening, just like the child that is born and named at the end of the story. Somebody has to take care of the child. The child is a promise of better times but somebody must first take the responsibility for bringing it up.

There are signs that we are beginning to see the end of the period of complete turmoil in Africa. In South Africa, nobody ever imagined that apartheid was going to end without some really terrible violence and the sort of chaos which Nadine Gordimer described in her novel *July's People*. We are not completely out of the woods in South Africa but the transition which has happened is almost miraculous. In the same way, the situation in Nigeria just seemed to get worse each day and then suddenly, without any apparent plan or reason—it seemed like providential intervention—people, like General Sani Abacha, began to die. It was as if some superior force was clearing the deck to give us another go. We've been given the opportunity of seeing exactly where we could end if we went down a certain road. We are no longer in the mood for thinking that an Idi Amin can only happen far away in Uganda or some such place. We now know that it can happen in Nigeria. This is something we had not considered even ten years ago. Then I would have said things could go wrong, but not as wrong as this. So I think that history does teach us. We're not good students, but in the end we do pick up pieces here and there. This is the hope, the only hope, perhaps.

I spoke at the World Bank in 1998 and one of the things I proposed was the cancellation of the Third World's debts if that can be managed. But I don't think that we in Africa should pin all our hope on what other people should, might, or might not do. I think we should get on with thinking about what we can do for ourselves, which may not be as dramatic but I think in the end might be more effective. We must put our house in order, guard against the emergence of undemocratic governments of all kinds and against the fascination that some people have for the so-called strong man.

Obviously things like the eradication of certain diseases such as malaria, which takes such a terrible toll, would be good. People don't talk about it any more, but that really would be a terrific thing to happen.

Hunger in Africa is due partly to sheer bad luck, to terrible droughts which have always happened and play a part in folk tales. But they can be handled with good technology and management and these are issues which should also be addressed. We are not addressing them because we are completely overwhelmed by the excesses of bad government, which lie like a curse on the continent.

The end of the Cold War should help. The powerful nations of the world have simply lost interest in Africa. They are no longer rushing around trying to find out who is a communist and who is not. This almost neglect of Africa, compared to what we had before, is a blessing. And I am optimistic about the new president in Nigeria, Olusegun Obasanjo. He was the one who handed over power to the civilians in 1979 and then went into retirement. Then providence came in and made him a victim of dictatorship. He's lucky to be alive. So nobody can tell him what it means to live under a despot. He's had special training, as it were, for this role as Head of State.

Given that fact, and given the readiness of Nigeria for change, and given the talent (Nigerians are extremely talented) and given the resources of Nigeria (oil, agriculture) and given the resources of their past (Nigeria has long been a maker of the arts) and given their special place in Africa (a quarter of the population of Africa is in that one little corner of the continent)—there are so many good signs. The signs were there all along but they have to be managed well. With this manager, Obasanjo, I think the chances are better than they have been for some time.

At the beginning of the twentieth-century William du Bois said that he hoped that the issue of the twentieth century would be Race, or the 'colour-line' as he called it. He was right, but he probably assumed that it would be solved by the end of the century. Well, it wasn't and one hopes that perhaps it will be in the twenty-first century.

Further Reading

Achebe, Chinua, *Things Fall Apart* (Oxford, 1958).
—— *No Longer at Ease* (Oxford, 1960).
—— *Arrow of God* (Oxford, 1964).
—— *A Man of the People* (Oxford, 1966).
—— *Home and Exile* (Oxford, 1999).

French Anderson

THE WEEK I SPEAK TO DR FRENCH ANDERSON THE FEDERAL BUREAU OF INVESTIGATION IS INVESTIGATING A DEATH THREAT DELIVERED TO HIS OFFICE. THE ANONYMOUS LETTER ARRIVED AT THE UNIVERSITY OF SOUTHERN CALIFORNIA SOON AFTER NEWS BROKE OF ANDERSON'S 'PRE-PROPOSAL' TO CONDUCT EXPERIMENTAL GENE-THERAPY ON FOETUSES IN THE WOMB. THE LETTER ENDS WITH THE MESSAGE: 'IT IS THE DUTY OF GOD'S DISCIPLES TO STRIKE YOU DEAD.'

Anderson, director of gene therapy at the university's medical school, handed the letter to university officials; it found its way to the postal inspectors, and thence to the FBI. His office has been cautious of strange packages since the late 1980s, when Anderson carried out the first gene-therapy treatments on human patients. The hate mail dropped off after a while, but Anderson's latest proposal—to inject engineered genes into foetuses in an attempt to correct heredi-tary ailments—has inspired hundreds of e-mails, faxes, and phone calls.

The writer of this latest death threat began his letter by paraphras-ing an 'action alert' put out by an American organization, the Council for Responsible Genetics. The council's board includes scientists like the eminent palaeontologist Stephen Jay Gould, 'whom I have known for years,' says French. Nonetheless, he is clearly unhappy about the fact that the threatening letter draws heavily on material originally published by the council.

For the CRG did not mince its words about Anderson's proposal. Under the headline 'Say No to Designer Children!!!' the council's 'action alert' dramatically announced 'This is it. This is how it begins.' It warned that Anderson was one of a group of scientists who 'favor a future in which the human race would be "improved" through genetic engineering'.

'If this first proposal is accepted, how much longer will it be before difference becomes defect, and any child who doesn't measure up to some arbitrary standard of health, behavior, or physique is seen as flawed?' the council asked. 'Do we want a future in which babies are produced according to genetic recipes?'

This kind of rhetoric, says Anderson, is misleading. It ignores the terrible suffering children with hereditary ailments are having to live through and presents, in emotive language, only one aspect of the argument about the genetic engineering of children.

In 1998 Anderson made a submission to the Recombinant DNA Advisory Committee (RAC) of the National Institutes of Health, the US committee that considers and can give the green light to sensitive genetic research. In his submission Anderson outlined a programme of gene therapy on foetuses in the womb designed to treat two conditions—ADA deficiency disease and alpha-thalassaemia.

Children born with ADA deficiency have no immune system. Until recently, and the advent of drug treatment, they were fated to live as 'bubble children'—in an infection-free plastic chamber. Foetuses with the blood disorder alpha-thalassaemia, meanwhile, do not produce haemoglobin. Mothers are forced to abort before their pregnancy reaches 24 weeks and the dying foetus becomes toxic to the mother.

Anderson has been at the forefront of administering experimental gene-therapy treatment to children suffering from ADA deficiency, which is also known as Severe Combined Immuno-Deficiency Disease. In 1990, with a team of scientists from the National Institutes of Health, he led a gene experiment on a young ADA sufferer. Since then more than 3,000 patients have received gene-engineered cells in the hope of replacing defective or missing genes. But the results have been inconclusive. There is no evidence yet of a miracle cure, and the research has left many questions unanswered.

Anderson keeps in close touch with the families of two young women, Ashanti DeSilva and Cindy Cutshall, whose treatment began in the early 1990s. 'Individual patients have been helped but there is not a high enough percentage of cells that get corrected to really be effective,' he admits.

Sadly, most of the patients treated so far had terminal cancer and have died. One worrying possibility is that the genetic experimentation itself may precipitate cancer—though this has not yet been proved in either humans or monkeys, 'so the risk is clearly not high,' says Anderson. While several hundred patients are still alive, they do not constitute a large enough sample to make it possible for doctors to be certain of the possible long-term side-effects of the gene therapy.

Some American scientists regard gene therapy on the foetus— when cells are dividing at a rapid rate—as a Holy Grail. If the therapy works in the womb, in theory, children could be born healthy and whole, and never need another medical treatment for their genetic disease in their lives.

Anderson's other proposal, to treat foetuses with the blood disorder alpha-thalassaemia—foetuses never expected to survive long enough even to be born—has aroused equal controversy. Its reception has been a painful lesson in medical ethics, not to mention in the politics of abortion in North America. Critics have agonized about the risk that partially successful treatment might produce a severely damaged child, who would live only a few weeks, possibly in great pain.

There is a second ugly issue too: the image of a medical team poring over an aborted foetus to check whether the experimental therapy actually worked. Anderson says it was an RAC member who suggested the 'stage-zero' study; to take a mother who planned to terminate her pregnancy anyway, carry out the gene therapy at 18–20 weeks and then evaluate the aborted foetus at 24 weeks. 'My response was that logically I could see it would make sense, but, personally I would have great difficulty doing it,' he says.

'Since then, as this discussion has evolved, I have suggested going ahead with treatment, then testing the foetus at 24 weeks. If we cure the foetus, then it shouldn't be aborted, but allowed to grow to term. But if we are only partially successful, then—the abortion should be performed.'

It sounds like an almost unbearable scenario for the parents who would be involved in such decisions, see-sawing between hope and despair. But parents, says Anderson, 'are extraordinarily supportive of doing things that will help not just their own future children, but will also help future mothers. We would want to insist on having outside advocates for the family to turn to, independent of our research teams so that they could really communicate about these issues . . . What one has to avoid is false hope, and it is very easy for desperate couples to develop a false hope.'

Anderson says his team submitted 180 pages to the RAC outlining potential problems with the research back in July 1998; this report was sent to twenty-seven individual reviewers, who weighed in with an inch-thick list of observations. He was not asking for approval, he told committee members, but was only opening a public debate.

And it is this debate he feels groups like the Council for Responsible Genetics should contribute to. So far he feels that the council's rhetoric has sometimes obfuscated rather than clarified his research. For instance, the CRG led those who assumed that Anderson's work was opening the door to germ-line transfer—whereby an altered gene can be passed down to future generations in sperm or egg. But this is not necessarily true, says Anderson. Rather, he is pursuing research into somatic gene therapy for individual children, whereby genetic changes are made only within somatic cells in the body rather than in sperm, egg, or embryonic cells. Nonetheless, Anderson accepts that there is a risk that the therapy could have unforeseen consequences.

The plan to treat ADA deficiency in the womb, for instance, involves injecting a stretch of DNA containing the gene for the ADA enzyme into a foetus during the second trimester of pregnancy. Experts agree that there is a risk of the new gene being absorbed not just by somatic cells but by germ cells in the eggs and sperm, which would mean that it could be passed down to future generations. Parents who decide on the procedure might, in effect, be making genetic choices not just for one child but for grandchildren, great-grandchildren, great-great-grandchildren and so on.

Whether or not Anderson's research eventually gets the go-ahead will hinge in part on animal studies to gauge the level of risk. Beyond that, he says, there is the question of what degree of risk is acceptable, and that is why he is seeking to force a discussion now. 'Most people would feel that the risk-benefit ratio is perfectly acceptable if the choice is between a dead foetus and a healthy individual who could have his/her own family, with a one in a million chance of carrying an engineered gene, and a 50–50 chance that such a gene will be beneficial,' he says. 'On the other hand, if the chances of an engineered gene getting into the germ-line are 50 per cent, then I certainly would not go forward.'

For all his caveats and cautions, however, there is a very real distinction between Anderson and many others in his field. He says that any deliberate germ line transfer is a decade or more away from satisfying medical and ethical criteria that it is safe and reliable. But, he also argues that once those criteria can be met, then germ-line transfer should take its place as a medical option for physician and patient despite the fact that it will affect more than one generation.

His critics, and much of current medical and public opinion, it appears, maintain a blanket 'no' to meddling with the human gene pool. They argue that there are other options enabling people to have children free from certain inherited conditions—adoption, for instance.

'I feel it would be unethical to do an intentional germ-line gene transfer at this time because we don't know enough scientifically to do it safely, we don't know enough medically to do it effectively, and we don't know enough ethically to do it wisely,' says Anderson.

'We are years away from satisfying any of those three criteria. But if you ask me "do I believe in germ-line gene therapy?," the answer is yes. When the time comes, then we must do it, because it is just plain human nature. What parent would willingly pass on lethal genes to their children if there's a safe, effective way of eliminating them?'

by Tim Cornwell

Gene Therapies

W. FRENCH ANDERSON

I predict that, by the year 2030, gene-based therapy will have revolutionized the practice of medicine. I define gene-based therapy as any therapy based on replacing or modifying the function of genes in the body.

The two primary branches will be gene therapy, whereby one or more genes are injected into the patient to treat a disease, and drug therapy, in which a drug is given to the patient to modify the expression of one or more genes in the body.

The Human Genome Project is now teaching us what all the human genes are and, in time, we will know what they do. We will rapidly develop the ability to screen for genetic defects or weaknesses. By 'weaknesses' I mean genes that do not function optimally for the environment in which the patient lives (whether that environment is stressful because of diet, radiation, toxins, or whatever). Such genes could therefore result in the patient developing a serious disease.

Once a defective or poorly functioning gene is discovered, we will have the ability to give a patient a more effective gene to replace the 'weak' one (gene therapy) or, if the gene is making a normal product but just too much or too little of it, a drug that can increase or decrease production from that gene. By the year 2030 I anticipate that there will be a gene-based treatment for every disease.

Drug therapy does not alter the human genome, but gene therapy does. Once we have the ability to give a patient any gene we want in order to treat a disease, then we will have the ability to give someone genes for any other purpose too. I fear that the downside of this powerful technology might be that eugenics will be practised on a scale far larger than any 'selective breeding' policy could accomplish.

From this point of view, society faces a massive threat. In the name of minor 'improvements' that we see as conveniences, we might start using human genetic engineering to attempt to 'improve' ourselves—and our children. Engineering the human germ-line would then result in permanent changes in the gene pool.

In the late twentieth century we have shown that we are ill-equipped to handle discrimination, even in its present forms. Already, at the close of the twentieth century the virulent evil of 'ethnic cleansing' is rampant in several countries. What would happen to us if we added intentional genetic enhancement to our world?

In 1998 Hollywood made a movie called *Gattacca*. In *Gattacca* only genetically cleansed individuals could hold good jobs; 'love children', who were produced by natural means, were relegated to the poorest positions.

I believe that the only protection for our society to prevent us tumbling down a slippery slope to a *Gattacca*-style society is to insert clear stopping points along the route. And the only means to do so is to develop an informed society which recognizes the dangers of genetic engineering and prevents misuses of the technology before it is too late.

We cannot dictate to society 100 years hence what it should do. The people of the future will care as little about our opinion as we care about long-forgotten mandates from our nineteenth-century predecessors. But it is our duty to go into the era of human genetic engineering in as responsible a manner as possible. This means that we should use gene therapy for no other purpose than the treatment of serious disease, no matter how tempting it might be to try to 'improve' ourselves with this powerful new technology.

Further Reading

Anderson, W. French, 'Human Gene Therapy: Why Draw a Line?', *J. Med. and Phil.* 14 (1989), 681–93.

—— 'Genetics and Human Malleability', *The Hastings Center Report*, 20 (1990), 21–4.

—— 'Gene Therapy', *Scientific American*, 273 (1995), 124–8.
—— *A New Front in the Battle Against Disease* (Los Angeles, in press, 1999).

Noam Chomsky

NOAM CHOMSKY IS NOT HAPPY WITH THE IDEA HE IS FAMOUS. PERHAPS 'INFAMOUS,' HE CONCEDES. OUTSIDE ACADEME, HE IS MOSTLY INVOLVED WITH POLITICAL ACTIVISTS, PEOPLE WHO DO NOT AUTOMATICALLY CONFER ON HIM ANY PARTICULAR HONOUR. HE TRIES TO AVOID THE PRESTIGIOUS LECTURE TRAIL. 'I DON'T THINK FAME IS EXACTLY THE WORD.'

But whatever it is that allows him to pack out auditoriums across the globe, that makes him a hit with alternative youth groups, hero-worshipped by some academics and politicians, violently attacked by others; whatever it is, it makes constant demands.

As he snatches a few extra minutes before our interview to deal with graduate application papers his secretary at the Massachusetts Institute of Technology, in Cambridge, Massachusetts, where he has been a professor for forty-five years, is reassuring. 'He'll be fine,' she says. 'He's in a good mood today.'

Actually, Chomsky apologizes, with a diffident smile, he may look 'a little haggard'. He was up until three o'clock this morning writing. He has already conducted one interview—with a group of high school kids. He has several manuscripts awaiting attention. The work is relentless. It takes him three or four hours a day to deal with e-mail. Other mail takes another couple of hours. Then there are requests for interviews, talks to prepare, departmental administration to address. He spends a lot of time on aeroplanes and fills it by writing, or thinking about future lectures. 'I live on a kind of survival strategy,' he says. 'See if I can make it to tomorrow. There are always deadlines pressing.'

Chomsky is prolific—he has published about seventy-five books, hundreds of articles, and thousands of letters. What really marks him out, and demands every waking minute, is the fact that he is prolific in so many different areas. Expert in subjects from linguistics to philosophy, history to mathematics, he describes his life as one of 'multiple personalities'.

But it is as a scholar of linguistics and as a political activist and scourge of the American authorities that he is best known.

In linguistics, he thought up the concept of a 'universal grammar', arguing that all children are born with a fixed set of mental rules for grammar—which enables them to make up sentences they have never heard before. For him, this explains how children acquire the complex rules and rhythms of language so easily; they already 'know' these rules as part of their biological endowment. Language is an essential component of the mind.

Many argue he has 'revolutionized' linguistics. It is certainly rare to read about the subject without his name at least cropping up. Others argue that he has focused on the wrong area. How does exploring ways in which language is innate help those for whom it is a problem? He is also criticized for concentrating too much on abstract theory, rather than on direct observation of how, for example, real children's speech develops.

He takes issue with this last criticism, pointing out that, although his work is not done in the lab, his belief that knowledge of language is something within us is supported by the findings of leading experimenters in child language, with some of whom he has worked closely for years.

While his work in linguistics has often put him at the heart of intellectual debates his political work has more than once had him thrown into jail. In the early 1960s he made a conscious decision to jeopardize a comfortable life of research and family by becoming actively involved in protests against the Vietnam war. Several nights a week he would address small, hostile gatherings. He recalls the first attempt to have a public demonstration against the war—in October 1965, on the Boston Common, the standard meeting place. 'I was supposed to be one of the speakers, but the mobs were so hostile that none of us could say an audible word. The only reason we weren't killed, I suppose, was that there were hundreds of police—who didn't like what we were saying one bit, but didn't want to see anyone murdered on the Common.'

The next time they tried was in April 1966—in a downtown church. Again the mobs turned out. Chomsky recalls standing next to a police captain at the church door. 'He was hit in the face with a tomato and the place was quickly cleared.'

Similar mishaps befell his wife and daughters in a women's demonstration. He comments drily: 'It wasn't until late 1966 that one could hold public demonstrations against the war without violent attacks in Boston, the Athens of America. There has never been any report of this in the various histories of the period. It's not the kind of thing nice people talk about.'

Since these early encounters he has consistently spoken up against American policies of violence and repression and the propaganda that accompany them. Propaganda, he argues, 'is to democracy what violence is to totalitarianism'.

He keeps up to date with political developments across the world through newspapers and official documents, always stressing the importance of addressing issues where he can make a difference.

He has tenaciously criticized American intervention in Indochina, Latin America and East Timor, the island fighting for independence from Indonesia (where he says the USA 'gave massive support to near genocidal massacres'). He despises the superpower mentality which chooses to intervene only in ways and in conflicts which suit itself.

America and Britain, he says, justify their bombing of Iraq in the 1990s by describing Saddam Hussein as a monster who gassed his own people. But they fail to add three words, 'with our support'. 'The United States and Britain supported Hussein in the full knowledge that he gassed his own people,' he says, referring to their sale of weapons to Iraq in the 1980's conflict with Iran.

Recently Chomsky has concentrated on attacking 'neoliberalism,' by which democratic principles are exploited in order to subordinate people to the profit needs of a handful of private interests. Among many examples, he includes Washington's 'crusade for democracy' in Latin America.

Chomsky cites Thomas Carothers, the leading academic scholar and State department official, on the Reaganite programme of 'exporting democracy' to countries like El Salvador and Nicaragua. Chomsky quotes Carothers' admission that the USA adopted 'only limited, top-down' forms of democracy, 'that did not risk upsetting the traditional structures of power—with which the US had long been allied.' This even when Latin American governments had a long record of abuse of human rights.

That the two sides of his life are distinct is something he insists on. His work in linguistics, he says, takes research, while his political ideas just demand a bit of thought. Take, for example, scares over how the 'baby boom' generation is to be cared for when it retires.

'It takes five minutes thought to see that that problem was dealt with when they were children,' he says. 'Anyone who will have to be cared for when they are 70 and when they are 90 also had to be cared for when they were aged 0 and when they were aged 20. The US didn't collapse under that burden. And now it's a much richer country. Why is a much richer country going to fall apart when it has to deal with a problem it already handled? This isn't quantum physics. Any school student can figure this out. But what we read and what we hear rammed into your head over and over is that the situation is at crisis point.'

This he calls 'good propaganda'. 'Good propaganda is to make people focus on something that is going to motivate them to do what you want,' he says. 'In this case, turn their money over to Wall Street, instead of giving them enough of a framework so that they can see that this is the wrong response.'

A slight man, now 70, he speaks so hesitantly that it can be hard to detect the bite behind his words. But they still hit home.

In fact, Chomsky inspires admiration to the point of awe. Linguist Raphael Salkie says he is 'brilliant' in linguistics, unpretentious and accessible. 'A friend said he got a sensual pleasure reading his political writings,' he says. 'I feel the same thing. I feel all the kinds of illusions and stupidity that get me confused are swept away and I get a beautiful clarity and understanding.'

Neil Smith, author of *Why Chomsky is Important*, says: 'He's of an intelligence which makes other people pale into insignificance. One has the feeling of being in the presence of genius.' He attributes Chomsky's 'fanaticism' to the desire 'to be seen to have done something decent with his life' but says he is 'a very human genius', boosted by a secure family life.

But, like all famous people, he also has his detractors, people like philosopher Daniel Dennett who accuses him of making linguistics such a nasty field that many academics now want to avoid it. Geoffrey Sampson, reader in computer science at the University of Sussex and a vociferous Chomsky critic asserts that 'most of his ideas about language are mistaken' and 'his picture of human nature is a very unattractive one.'

Born in Philadelphia to first generation Jewish immigrants from Eastern Europe, Chomsky's father was head of a Hebrew college. He worked on Semitic linguistics. His mother endowed him with a 'general concern about social issues'. There was always much talk about such issues at home and he had a number of relatives who were involved in radical Jewish politics.

Chomsky says he started questioning American society when he was 4 years old and saw people coming to the door selling rags, trying to survive.

He was educated from the age of 18 months in a progressive Deweyite school, which emphasized creativity and eschewed competitiveness, and which he remembers vividly as 'a very exciting experience'. His brother, now a doctor, went to the same school and their wives—Chomsky's wife, linguist Carol Schatz, and his sister-in-law, a radical lawyer—came from similar backgrounds.

At the age of 10 he wrote an article for the school newspaper on the Spanish Civil War, lamenting the rise of fascism. By the age of 12 he was hanging around anarchist offices and becoming involved in youth activities, particularly in Zionism. However, it was only when he reached high school that he discovered he was 'a good student, all As'—and that he hated the competitive, regimented aspects of normal school life.

As an undergraduate at the University of Pennsylvania he remained uncertain about academic pursuits. Commuting from home and teaching Hebrew in his spare time to earn his keep, he remained active in radical anarchist and left-wing politics. With no particular academic ambitions, his plan was to drop out and live in a kibbutz, where he later did spend some time before his career took off in earnest. But he was persuaded to stay at Pennsylvania in linguistics classes and to further a growing interest in politics and philosophy. 'It wasn't a first rate university at the time but it had some very good people scattered around,' he says. 'A lot of students who were there ended up with a strange collection of interests because they gravitated to whoever were the good teachers.'

A graduate fellowship at Harvard followed, and then a research post at the Massachusetts Institute of Technology in 1955, when he was also awarded a Ph.D for his thesis on 'Transformational Grammar', a combination of theory and analysis of English. Two years later he wrote *Syntactic Structures*, which made his name as a linguist, and within four years had been appointed to a chair.

Beginning work at MIT in a military lab—'I was the first person they remembered who had ever refused to get clearance'—he nonetheless said this was the freest system around. 'When MIT was funded maybe 90 per cent by the military it had no constraints on what it should do,' he says. 'As it has moved from the Pentagon to corporate funding there are more and more constraints.'

Yet he is scathing about intellectuals, accusing them of pandering to the establishment. 'If you are not subordinate to power you rarely make it through the system,' he says. This does not necessarily mean suppressing your own views. 'You just internalize—internalize the values so that [as George Orwell said] there are certain things it wouldn't do to say or even to think.'

Chomsky does not necessarily want everyone to agree with him. 'On the contrary, disagreement is much more interesting. But you have to at least accept some rules of discourse. Like rational arguments matter. Facts matter. And for a large part of the intellectual establishment they don't.'

He says he has been lucky at MIT because it is a science-based university and therefore naturally more subversive. 'Core education in the sciences is getting students to recognize that they are not supposed to respect authority. They are supposed to question and challenge and create good ideas. That comes to exactly the opposite of what most education is, which is mostly designed to instil obedience to authority and belief in power interests.'

There is nothing Chomsky would advise people to do that he would not do himself. He has only given up direct action and political organizing because he realized he was not that good at it. 'I can do much better helping people get their thoughts in order.' The aim is to get people to question, to think. It must take enormous self-confidence to find himself constantly going against the grain. 'I think it's a normal human endowment,' he says. 'Children have it. They are almost trained to doubt themselves.'

There is a knock at the door—he ignored the first, five minutes ago. Someone else clearly needs his time.

by Harriet Swain

Language Design

NOAM CHOMSKY

I t is hard to contemplate a request for predictions for the coming century without serious reservations. The record of prediction in human affairs has not been inspiring, even short-range; nor in the sciences. Understanding is thin apart from a few areas, and some crucial factors—such as human will—escape our intellectual grasp. Perhaps the most plausible prediction is that any prediction about serious matters is likely to be off the mark, except by accident.

One question that might be answered in the next century is whether humans are a kind of lethal mutation. The species appears in the last flick of an evolutionary eye, and has now achieved the capacity to destroy itself (and much else) by means ranging from weapons of mass destruction to environmental catastrophes. Perhaps it will find ways to contain its destructive impulses, and to address what may be ominous problems. A rational Martian spectator might not be sanguine about the prospects.

If the first question remains unanswered—the most that can be realistically hoped—generations to come have fascinating inquiries to pursue. To continue with the query about whether we are a kind of lethal mutation: why does that question arise specifically for humans? What properties of this curious species account for its unusual place in the biological order? Here we turn to questions about human higher mental faculties, about which little is understood—not surprisingly; similar questions are hard to answer for insects. It has been—or should have been—a truism in the modern era that 'the powers of sensation or perception and thought' are properties of 'a certain organized system of matter', that properties 'termed mental' are 'the result [of the] organical structure' of the brain and 'the human nervous system' generally (Joseph Priestley). But how the properties 'termed mental' emerge remains about as mysterious as it was 200 years ago. Some questions might be answerable by investigations that only a Mengele would pursue. Current technological progress may help overcome some of the ethical barriers with non-invasive experimentation. Considerable optimism has been expressed, but evidence is limited.

There is little doubt that the human language faculty is a core element of specific human nature. In this domain, a great deal has been learned recently, enough so that it is possible at least to pose, sometimes partially to answer, questions that could scarcely be contemplated only a few years ago. Thus, we can envision the prospect that, over an interesting range, human languages will be shown to be deducible from principles of (essentially) shared biological endowment by setting values for fixed options of variation, for example, for the place of a verb in a sentence. Implications for the study of language acquisition, use, and disability, and potentially the brain sciences, are rich, and being productively explored.

Recent work suggests more far-reaching possibilities. The minimal conditions on usability of language are that languages provide the means to express the thoughts we have with the available sensorimotor apparatus. One far-reaching possibility is that, in non-trivial respects, the language faculty approaches an optimal solution to these minimal design specifications (with 'optimality' characterized in natural computational terms) . If true, that would suggest interesting directions for the study of neural realization, and perhaps for the further investigation of the critical role of physical law and mathematical properties of complex systems in constraining the 'channel' within which natural selection proceeds.

What the work on optimal design of language seems to suggest is that language might be more like the appearance of familiar mathematical structures in nature such as shells of viruses or snowflakes than like prey/predator becoming faster to escape/catch one another. When the brain reached a certain state, some small change might have led to a reorganization of structure that included a (reasonably well-designed) language faculty. Maybe.

Such speculations are not completely without warrant. If they have some measure of validity, they should, in particular, advance understanding of fundamental human nature, though in ways that cannot be predicted with any confidence. Nevertheless, they still fall far short of classical problems that remain as mysterious as ever, for example, problems of will and choice—perhaps, we may discover, for reasons rooted in our cognitive nature, a conclusion that should come as no great surprise to those who take for granted that humans are part of the organic world.

Further Reading

Chomsky, Noam, *Language and Mind* (New York, 1968).
—— *Reflections on Language* (London, 1976).
—— *Rules and Representations* (Oxford, 1980).
—— *Language and Problems of Knowledge* (Cambridge, Mass., 1988).
Radford, A., *Transformational Syntax: A Student's Guide to Chomsky's Extended Standard Theory* (Cambridge, 1981).
—— ed. Smith, N. V., *Chomsky: Ideas and Ideals* (Cambridge, 1999) (essays and writings).
—— *Syntax: A Minimalist Introduction* (Cambridge, 1997).

Arthur C. Clarke

T O THOSE WHO KNOW HIM AS A LEADING SCIENCE–FICTION WRITER AND FUTUROLOGIST—THE AUTHOR OF *2001: A SPACE ODYSSEY*—ARTHUR C. CLARKE'S LATEST BOOK, A LIFETIME'S SELECTED ESSAYS ENTITLED *GREETINGS, CARBON-BASED BIPEDS!*, OPENS WITH A SURPRISING BUT CHARACTERISTIC DEDICATION: 'TO VICE-CHANCELLOR PROFESSOR C. PATUWATHAVITHANE, KILLED WHILE SERVING HIS STUDENTS, AND TO THE CHILDREN OF SRI LANKA'S LOST GENERATION, REMEMBERED ONLY BY THOSE WHO LOVED THEM.' THE PROFESSOR, CLARKE'S FRIEND, WAS ONE OF THE MANY VICTIMS OF SRI LANKA'S VICIOUS CIVIL WAR.

It is surprising—because Sri Lanka, though Clarke's island base for nearly half a century, for most of his readers does not appear to have impinged much on his prolific, best-selling fiction and non-fiction since the 1950s. It is characteristic—because it hints at how much the writer feels for his adopted country, despite the violence that has disfigured it in recent times.

Clarke is chancellor of the University of Moratuwa, half an hour's drive down the coast from his home in the Sri Lankan capital Colombo. He has stuck at the job throughout the political turbulence of the late 1980s and the 1990s. At one point he even considered having to leave the country for good. Opposite the university, bristling with satellite dishes, stands the Arthur C. Clarke Institute for Modern Technologies. It was originally to have been the Developing World Communications Centre—combining two of Clarke's long-term concerns, development and communication—in which he invested some tens of thousands of dollars awarded him as part of a Marconi International Fellowship in 1982. Then, in 1984, the Sri Lankan government incorporated it under its new name, with Clarke as its 'patron'.

Among its activities has been the development of locally designed control systems for traffic lights in Colombo: part of the centre's goal to redirect Sri Lanka's craze for imported technology.

It is generally accepted—by scientists as well as by politicians, media moguls, and journalists—that Clarke is 'the father of satellite communications,' on the strength of his far-sighted article, 'Extra-terrestrial relays,' published in *Wireless World* in October 1945. That paper proposed the concept of space stations circling the earth at a distance of about 36,000 kilometres in geostationary orbits, so that they would remain fixed relative to the earth's surface and thus able to act as transmitters of radio waves.

Clarke himself, deferring to the engineers who made the concept a reality, prefers to style himself 'godfather' of satellite communications. 'If I hadn't written that paper in October '45, ten people would have done it the next year, you see.' He goes on to dismiss any grand claims—frequently aired by others—for the originality of his scientific ideas, which he calls 'only propaganda on space'. 'I may have accelerated the conquest of space by 15 minutes or so,' he says modestly.

Comparing himself with the scientists he really looks up to, many of whom have been his friends—men such as J. B. S. Haldane, Wernher von Braun, and Luis Alvarez—Clarke is undoubtedly accurate in his self-assessment. But, unlike the vast majority of scientists, he has the power to convey the wonder of the universe to almost anyone. Stanley Kubrick, film director, his collaborator on the film *2001*, once remarked: 'Arthur's ability to impart poignancy to a dying ocean or an intelligent vapour is unique.'

Yet, getting his work right scientifically is of key importance to Clarke. One of his novels, *The Fountains of Paradise*, envisions an energy-efficient 'space elevator' or 'orbital tower' to replace rockets, with cables rising from the top of a famous peak in Sri Lanka, near the equator (as required by physics), and anchored in space by an orbital space station at around the same height as geostationary satellites.

This fantastic proposal was not original to Clarke; it was first conceived in 1960 by a Russian engineer, Yuri Artsutanov (who called it a 'heavenly funicular'), and was then independently reinvented at least four times by US scientists in the 1960s and 1970s. Alerted by his network of scientific friends, Clarke dramatized the idea from the technical literature and published his novel in 1979. When it reappeared as part of the novel *3001: The Final Odyssey*, and was described as a 'cylindrical tower' in a London *Sunday Times* review, an Indian scientist at the Nehru Planetarium who read the review claimed in print that the structure was impossible to build, that the height limit for a tower was only 10 km rather than the 36,000 km of the novel, which the scientist described as not science fiction but an 'outlandish fantasy'—to believe otherwise, he said, was 'like believing that matter can travel faster than light.' Another scientist supported the first and soon, to quote an Internet magazine, 'the media was abuzz with the fall of a giant like Clarke'.

Contacted by the magazine (by old-fashioned phone), a stung Clarke pointed out that his critics had completely misunderstood the structure of the space elevator and had also overlooked the strength of new carbon fibres. He e-mailed back: 'Suffice to say that the two recent Nobel winners (in chemistry), Richard Smalley and Harold Kroto, have both complimented me for my accurate prediction of the material—C_{60}—that will make the tower possible—and which has three times the tensile strength necessary.'

Clarke's e-mail was followed by one from a US Airforce aerospace engineer, an expert on the space elevator, who commented that the proposed structure was not a building, 'not a tower in compression; it is a cable in tension. Such a cable can be hung to any length without buckling. . . . A tapered hanging cable from geostationary orbit to the earth could be built . . . with carbon nanotubes, based on the theoretical strength of buckminsterfullerene, with a taper ratio of only the order of ten.'

Though the space elevator is scientifically feasible, it is unlikely to be built for a century or two (though a 1999 conference held by the

American space agency, Nasa, visualized a construction date as early as 2060). Clarke's short-run preoccupation, which he is willing to fund a major British university to research, is of more immediate importance. 'I would like to settle what is probably the greatest scandal in the history of science,' he says, 'what I believe may be a technology revolution comparable to the discovery of fire. I refer of course to so-called "Cold Fusion"—an unfortunate name because many of the devices concerned have nothing to do with fusion and are certainly not cold.'

Clarke has been following developments in many countries since 1989, when the original claim of 'cold fusion'—that atomic nuclei can be fused at room temperature instead of at the temperature inside the sun or in a hydrogen bomb—caused an unprecedented global scientific furore, before being discredited.

The scientific orthodoxy remains hot fusion research, consuming billions of dollars without much sign of useful progress. Clarke maintains: 'Although many of the claims made for "cold fusion" are certainly delusions, or even frauds, what remains is enough to convince any unbiased observer that some new sources of energy are being tapped. The mechanism may be "sonofusion" induced by microcavitation in liquids, or even the inconceivably vast energy of quantum fluctuations, which is known to exist and has been detected in numerous experiments—or it may be something completely new.'

He likes to quote the remark of Richard Feynman, physics Nobel laureate, that the zero point energy from quantum fluctuations in a volume of space no larger than a coffee mug is 'enough to boil all the oceans of the world.' But his enthusiasm, publicized by the BBC and national newspapers, has attracted some distinguished critics, such as the Nobel prize-winning physicist Philip W. Anderson, who wrote that 'cold fusion is a good paradigm for what really happens when we have "socially constructed" science'.

While scientific errors are rare in Clarke's writing, he has many times been wrong about technological progress. 'Often,' he says disarmingly in the preface to his book of essays, 'it has been more interesting to see where (and why) I went wrong, than where I happened to be right.'

He comments: 'One tends to be over-optimistic in the short run and under-optimistic in the long run, because we can only extrapolate linearly and progress is always an exponential curve. Sooner or later the exponential curve crosses the linear extrapolation.' Maybe the best-known example is the moon landing, which, in the 1940s, he never really expected to see in his lifetime—but then who could possibly have predicted the space race between the superpowers in the 1960s?

It is significant that Clarke, of all people, first learnt of the criticism of the concept of a space elevator in his novel *3001*, not via the Internet, but by telephone. In *2001*, he described with prescient excitement the journey of Dr Heywood Floyd, key scientist and troubleshooter, on his way by space shuttle to the moon, routinely scanning the latest reports in earth's electronic newspapers on his 'foolscap-sized newspad'. 'Switching to the display unit's short-term memory, he would hold the front page while he quickly searched the headlines. . . . Each had its own two-digit reference; when he punched that, the postage-stamp-sized rectangle would expand until it neatly filled the screen, and he could read it with comfort.'

But, having neatly prefigured the electronic media of the late 1990s, in the next paragraph Clarke wrote: 'The text was updated automatically on every hour; even if one read only the English versions one could spend an entire lifetime doing nothing but absorb the ever-changing flow of information from the news satellites.' That was written in 1968.

In 1997 he told me: 'I have not looked at the net, not even my own websites'—there are several devoted to his work—'because you get bogged down in a mess of information. It's something I can live without. Any spare time I have now is for working on the essays.' Then, a moment later, he said, brightening at the thought: 'When *Surveyor* [a spacecraft now in Mars orbit] starts doing things on Mars, I'll certainly be clocking into the JPL site' (he has many friends at the

Jet Propulsion Laboratory in California). 'The Internet's too slow here. If I got faster access, I'm sure I'd use it much more, though I'll probably regret that. It's becoming far more than I ever imagined. It could very well mean the end of the nation state, and the development of—what's the word—tribes, who only owe minimal allegiance to their country of birth.'

Does he welcome this possible future? Surely not, if one judges from a piece he wrote about one of his heroes, author-philosopher Olaf Stapledon: 'Today's couch potatoes, surfing the channels with their remote controls, may be the precursors of Stapledon's still more sedentary "Great Brains", encased in their concrete igloos yet free to roam the world via their mobile sense organs.' He tells me: 'It's the fractal future. Although everybody is ultimately connected to everybody else, the branches of the fractal universe are so many orders of magnitude away from each other, that really nobody knows anyone else. We will have no common universe of discourse. You and I can talk together because we know when I mention poets and so on who they are. But in another generation this sort of conversation may be impossible because everyone will have an enormously wide but shallow background of experience that overlaps only a few per cent. As time goes on, all the great classics—who will even know what they are? Who will even know who Shakespeare is in 1,000 years' time? It's a terrible tragedy, isn't it? I don't know the answer.'

He summarizes his ambivalent feelings: 'I'm in favour of communication developments—everybody will be in touch with the whole universe. But are we just going to be swamped in a sea of noise? Or will it be all right on the night?' Then he asserts: 'I like the idea of cultural diversity—that's one reason I find it fascinating to be here in Sri Lanka. I'd hate to see us all wearing grey-flannel suits or'—he pauses briefly and grins—'grey-flannel sarongs'.

Had Clarke remained in his native Britain instead of relocating to Ceylon in the 1950s, it is a certainty that he would never have donned his customary sarong, and a fair bet that he would have been swamped by his British and American fame and fans. But what initially lured him to Serendip was the warm climate and his childhood love, the sea, with the endless thrill of underwater diving and exploration. (He quickly set up a small company, now called Underwater Safaris, and in the early 1960s, with his partner Mike Wilson, discovered and salvaged treasure from a sunken Mughal galleon.)

Only later did he discover eastern mystery, the island's extraordinarily rich cultural and religious traditions, that have infiltrated some of his books. He says frankly: 'I suppose I didn't even know this was a Buddhist country, when I first came here.'

Not long ago, a friend took what is probably Clarke's most famous story, 'The Nine Billion Names of God,' an ironic tale of two computer engineers hired by Tibetan monks to generate the nine billion names of God, in which the monks prove to be the wiser party, and gave it to the Dalai Lama, who found it 'particularly amusing'. 'I would be interested to meet him,' says Clarke, 'but I don't suppose we'd have much in common.' A few years earlier, Clarke lectured at the Pontifical Academy of Science in Rome and presented a copy of his collected scientific articles, *Ascent to Orbit*, to Pope John Paul II. Despite an absolute, not to say militant rejection of organized religion (and, almost equally, of pseudo-science, such as spoon bending, UFOs, and alien abductions), Clarke's writings are marked by a persistent search for God—among the stars, rather than within the human mind. His tone varies from the deadly serious to the bantering. (In *3001*, he has a doctor note that the dangerous practice of male circumcision was only finally abandoned when 'some unknown genius coined a slogan: "God designed us: circumcision is blasphemy".')

A Clarke scholar, Eric Rabkin, writes that 'he really does seem to want to believe there's something higher and that it cares for us'. The comment is cited by Clarke's authorized biographer, Neil McAleer, who glosses it: 'Such a quest for faith appears throughout much of Arthur Clarke's fiction, and this continuing search, which takes many forms in his work, is one of the foremost reasons that his novels have permanent value.'

'Quest for faith': the biographee is uncomfortable when I remind him of the phrase. 'Well, I have a curiosity,' he says, 'but there's certainly no question of a burning desire to have all the answers and be saved, which some people have. I've only a philosophical interest in all these things. Reincarnation, which everyone here in Sri Lanka believes in, is a fascinating idea, but I'm not sure it's an attractive one. I'd like to feel that when you've had your time, that's it: the carbon's used for some other biped, or multiped, as the case may be.'

A few minutes later, he adds, with puckish quirkiness, 'My problem with reincarnation is that I don't see any mechanism that would make it work. What are the input–output devices, what's the storage medium?'

Perhaps he encapsulated his underlying attitude best in an essay written in 1989, entitled 'Credo'. He wrote: 'Men have debated the problems of existence for thousands of years—and *that* is precisely why I am sceptical about most of the answers. One of the great lessons of modern science is that millennia are only moments. It is not likely that ultimate questions will be settled in such a short time, or that we will really know much about the universe while we are crawling around in the playpen of the solar system.'

If, a millennium from now, when 3001 rolls in, Arthur C. Clarke is still remembered, it will most likely be for the force of his conviction that the future of humanity would be cosmic.

by Andrew Robinson

2099 . . . The Beginning of History

ARTHUR C. CLARKE

Despite all claims to the contrary, no one can predict the future and I have always resisted all attempts to label me a 'prophet': I prefer 'extrapolator'.

What I have tried to do, at least in my non-fiction, is outline possible 'futures'—at the same time pointing out that totally unexpected inventions or events can make any forecasts absurd after a very few years. The classic example is the statement in the late 1940s by the then chairman of IBM that the world market for computers was about five (or was it six?). I have more in my own office and they are breeding like rabbits . . .

But perhaps I'm in no position to criticize Thomas Watson, Snr. In my short story 'Transit of Earth' (1971) I put the first Mars landing in 1994; now we'll be lucky if we make it by 2010. Still, I take a modest pride in the fact that communications satellites are placed exactly where I suggested in 1945. And the chapter 'The Century Syndrome' in my 1990 novel *The Ghost from the Grand Banks*, may well have been the first account, outside the technical literature, of the now-dreaded Millennium Bug—its cause and its cure.

So, the predictions I make here come with a health warning. Some are already scheduled (the space missions, for example), but I believe all the other events could happen in the fullness of time.

Early in the twenty-first century the *Cassini* space probe will begin exploration of the planet Saturn's moons and rings while the *Galileo* space probe continues surveying Jupiter and its sixteen moons.

Life beneath the ice-covered oceans of Jupiter's moon Europa will appear increasingly likely and before the century is halfway through large marine creatures will be discovered when the first robot probes drill through Europa's ice, revealing an entire new biota.

In 2003 [the American space agency] Nasa's robot *Surveyor* will be launched and a couple of years later it will send its first sample from the Red Planet back to Earth. Within another decade aerospace planes will enter service and Prince Harry will become the first member of the Royal Family to fly in space. He may even stop off at the Hilton Orbiter hotel, assembled out of the giant tanks from the Shuttle spacecraft [America's spacecraft of the 1980s], which had previously been allowed to fall back to Earth.

After 2020, when artificial intelligence reaches human levels, there will be two intelligent species on Planet Earth, one evolving far more rapidly than biology would ever permit. Interstellar probes carrying artificial intelligences will be launched towards the nearer stars.

A year later the first humans will land on Mars and in 2057, the centennial of *Sputnik 1*, the first artificial satellite, launched by the Soviets, the dawn of the space age will be celebrated by humans not only on Earth, but on the Moon, Mars, Jupiter's moons Europa and Ganymede and on Saturn's largest moon Titan—as well as in orbit around Venus, Neptune, and Pluto.

The end of the century will see the development of a true 'space drive'—a propulsion system reacting against the structure of space-time. This will make the space rocket obsolete and permit velocities close to that of light. The first explorers will set off to nearby star systems that robot probes have already found promising and then history will truly begin.

Further Reading

Clarke, Arthur C., *Childhood's End* (New York and London, 1953) (novel).
—— *2001: A Space Odyssey* (New York and London, 1968) (novel).

—— *Ascent to Orbit: The Technical Writings of Arthur C. Clarke* (New York, 1984).

—— *Greetings, Carbon-Based Bipeds!: Collected Works, 1934–1998* (New York, 1999) (essays).

McAleer, Neil, *Odyssey: The Authorised Biography of Arthur C. Clarke* (London, 1992).

Paul Davies

PAUL DAVIES IS ANXIOUS TO EXPLAIN WHY HE IS HERE, BEING INTERVIEWED, IN A BARE BUT EXCLUSIVE TEMPORARY APARTMENT JUST OFF SLOANE SQUARE IN LONDON. HE IS IN ENORMOUS DEMAND, HE SAYS. HE GETS INVITED TO APPEAR ON TELEVISION, GIVE LECTURES, AND ADDRESS COMPANIES AS DIVERSE AS THE AUSTRALIAN RAF AND THE TAX OFFICE, SEVERAL TIMES A DAY, ALL OVER THE WORLD. HUNDREDS OF PEOPLE COME TO SEE HIM.

While he is fairly famous in Britain, both as a popular science writer and as a physicist with some rather spiritual views about the nature of the universe, in Australia, where he now lives, he gets recognized in the street and stopped in supermarkets. He finds it 'very gratifying to think people are so enthusiastic about a scientist.' He has met the Pope a couple of times and the Dalai Lama. 'A lot of my working time is spent scheduling my activities, sitting staring at a diary trying to work out how I can fit everything in,' he sighs.

Davies has always been interested in the big issues—the beginning of life, the nature of time, the (as the title of one of his books puts it) mind of God. 'As a teenager, I used to lie in bed wondering where do I come from, what will happen when I die and so on,' he says. 'I used to worry a lot about free will. Anything that was hidden interested me.'

What has always fascinated him is how large- and small-scale problems fit together. From early on in his career, he combined an interest in calculations at particle level with an interest in cosmology. His early work focused on the properties of atoms and molecules explored through quantum mechanics. Then he became interested in the asymmetry of time—the way in which fundamental laws of physics, making no distinction between one moment and the next, co-exist with time's arrow, pointing ever towards the future. Davies looked for the answer in cosmology and the arrow of time present in the whole universe, from the Big Bang to black holes.

An early Davies book was the co-authored monograph *Quantum Fields in Curved Space*, which became an important text in the field of quantum gravity. The book was an attempt to marry quantum theory with gravitation. This led him to reflect on what lay behind the complex web linking quantum laws to those of the cosmos.

More recently, he has looked at how the universe began, at relationships between science and religion and at those between physics and biology. He has written, too, about the nature of human consciousness, the question of how our subjective awareness of everything, colours, texture, pain, is formed.

The young Davies spent hours lying on his back in the garden with his home-made telescope, looking up at the stars and wondering. And in some ways, except for the moustache, there is still something about him of the bright child who has never stopped asking his weary parents 'But why?' All of which might explain his controversial interest in religion, a subject some of his colleagues feel is scarcely proper for a scientist.

Davies's belief that the universe is rational, that there is 'some sort of deep meaning to physical existence' has made him a leading member of what he calls an 'alternative academy' of scientists. This academy, he revealed a few years ago, is 'trying to construct a view of the world which is not religious in the conventional sense—but is more comforting than the bleak reductionism of most science over the past three hundred years.'

For despite being brought up in a moderately religious family, as a teenager Davies soon realized he was learning more about life's bigger mysteries from science than from the church. For twenty years he had almost nothing to do with organized religion and he has kept his four children well away from it. But as he started on the popular science trail he found that most of the questions at the end of his lectures were quasi-religious.

His own view has changed little since he was 16, he says. 'It has always been that the universe isn't ultimately absurd, that it is about something, that it is rational.' Opposed to the idea of a God as a sort of cosmic magician, he believes that the structure of the physical world and the existence of minds able to comprehend it are bound up together. But there is no magic about this. Science has already revealed many of the mysteries explained away in religious terms by previous generations and there is no reason, he says, why it should not eventually explain the rest.

In his book, *The Fifth Miracle: The Search for the Origin of Life*, Davies argues, unusually, that the key to life's beginning—one of the great unsolved mysteries of science—is information. While most writing on the subject examines the start of life as a bizarre chemical accident, he suggests, rather, that life was sparked by information being organized in some way, almost like a computer process. He also explores the theory that life did not begin as a chemical self-assembly in some watery medium on earth, nor did it arrive from space in the form of microbes, but that it began beneath the surface of the earth (and similarly inside Mars); that we live in an ingeniously biofriendly universe.

Relaxed and fluent, he could be explaining the workings of a dishwasher. Except that one reason Davies is so anxious to explain complicated ideas as clearly as possible is because of a belief in opening the eyes of others to the world's beauty. He has written: 'Nature, as revealed by science and mathematics, is altogether richer, more inspiring and more astonishing than our finest poets can portray.'

Life began for Davies fifty-three years ago in Charing Cross Hospital, North London. He was brought up in Hampstead Garden Suburb and Finchley and still retains a Hampstead accent, with just a trace of Oz. His father worked as a local government officer for the London borough of Camden and his mother, as well as bringing up three children, worked part-time for a secretarial agency. The early post-war years must have been a struggle, he thinks, for his parents.

But by the time he was 14 and the family was comfortably off, Davies knew he wanted to do physics, theoretical physics with a specialism in cosmology, to be precise. He asked his chemistry teacher how to become an astronomer. 'He gave me a booklet,' says Davies. 'But he said he didn't know anyone who had done it and didn't think it would pay very well.'

Books inspired him, particularly those of Fred Hoyle, best known for his support for the steady state theory of cosmology—which describes an eternal, infinite universe. Another influence was former prime minister Margaret Thatcher, who, as his local MP, presented him with a school science prize of an atlas of stars. Many years later, in 1995, she was also a judge of the £650,000 Templeton Prize—awarded annually, for 'progress in religion'.

By this time Davies had left Britain for Australia to become professor of mathematical physics at Adelaide, driven out, he says by the Thatcher government's cuts in funding to universities—what he has called 'the privatization of the universities by stealth'. Nonetheless, when Davies picked up the Templeton Prize, he took along his school atlas, and asked Thatcher to sign it.

It was at Cambridge in the 1970s, where he worked at the Institute of Theoretical Astronomy, that Davies discovered something which came easily and which turned him from being a moderately successful physicist into a celebrity. He had just scraped through English at school and never thought of himself as a writer. But he was asked by the journal *Physics Bulletin* to write about the nature of time. Interest from a publisher turned the article into a book, the magazine *New Scientist* asked him for regular pieces and the scientific journal *Nature* gave him a *News and Views* column. He was then asked to write a student text. 'With each step the level came down,' says Davies. 'By the end of the 1970s I was writing unashamedly popular books.'

Popular they may have been, but not necessarily with other scientists. Former colleagues recall feeling that he could not be serious about academic work and write for a general reader. Even Davies thought of his writing as 'parasitic on my research' and 'a bit of an embarrassment'.

All this changed when he arrived at Newcastle University at the age of 34 to take up a chair in theoretical physics. Physics as an academic subject had started its decline and universities were beginning to acknowledge that it was a good idea to communicate with the public. Davies began to receive plaudits for his writing. John Gribbin, a science writer and broadcaster who has known him more than twenty years, says Davies was ahead of his time in realizing that the more the public knew about science the more government funding was likely to be available for research.

But now, Davies, a man with a social conscience, also feels a duty to satisfy public demand. As well as the books—more than twenty-five at the last count—he averages one television or radio appearance per day. And for him it is so easy. He could write a book on black holes in two or three weeks, at the rate of a chapter a day, 'because it's all at my fingertips'. 'The point about popularising science is that ideas have their time,' says Davies, after complaining about journalists' failure to pick up on big science stories quickly enough. He gets irritated about this kind of mistake because he believes science offers people answers to life's big questions. He has never supported the idea of revelling in mystery.

by Harriet Swain

Three 'Origin' Mysteries

PAUL DAVIES

In the twenty-first century I should like to see the solution to at least one of the great outstanding 'origin' mysteries—by which I mean

- the origin of the universe
- the origin of life
- and the origin of human consciousness.

Remarkably, the first of these topics is the one on which most theoretical progress has been made so far. Physicists have the outline of a plausible mathematical theory explaining how a universe can originate from nothing. But without observational tests such mathematical models are of limited interest. Although we can expect major advances in our understanding of the very early universe from forthcoming astronomical discoveries, the originating cosmic event itself may lie forever beyond the scope of empirical enquiry.

By contrast, it may well prove possible to create life in the laboratory in the next few decades. At the practical level, that would open up the prospect of 'designer bioforms'—synthetic life produced by scientists to suit certain niches or fulfil commercial functions. For example, we might be able to make microbes that would thrive on some of the hostile planets in the outer solar system. Or bacteria that could clean up chemical or nuclear waste, or manufacture new materials. Already biochemists are genetically adapting existing life for similar purposes, but if we could synthesize new life from scratch, who knows what the applications might be?

More significantly for science, creating life 'in a test tube' would enable us to study the organizational principles that transform a mixture of non-living chemicals into a living cell. We could try to figure out how easy it was for Mother Nature to have attained the same result four billion years ago in the rough-and-ready conditions of the primeval earth, without fancy equipment or teams of trained chemists on hand to manipulate the reactions. It would also show whether the basic life processes are unique, or whether there are many alternative forms of life possible.

However, this research could be overtaken by events if we discover a second sample of life on Mars. But before drawing sweeping conclusions, it would be necessary to determine that Martian life emerged independently of terrestrial life, and isn't merely the product of planetary cross-contamination caused by the exchange of meteorites.

If we find that life has arisen twice within the solar system, it would be the greatest scientific discovery of all time. For example, it would point to a universe teeming with biological organizms, and offer the prospect of other intelligent beings inhabiting earthlike planets in the galaxy. It would also suggest that the laws of nature are ingeniously biofriendly, with all the philosophical implications that follow.

Human consciousness is a tougher nut to crack, since nobody knows what it is or how it is caused, except in the vaguest sense. I expect rapid progress in determining the so-called neural correlates of consciousness—what specific brain functions go with which sensations, feelings, etc. But the hard problem is to understand the subjective side of consciousness, the things that philosophers call *qualia*. For example, my inner experience of the colour red is quite distinct from that of green, or the sound of a bell, or the feel of water. Why do some electrical patterns, such as those in my brain, create *qualia*, while others, like those in the circuitry of my house, presumably do not?

These questions relate to the prospect of building thinking machines—intelligent computers that may also have some elements of consciousness. Could a man-made machine experience *qualia*, and tell us about them? The possibility may seem far-fetched, yet some researchers think they are already approaching the goal of artificial intelligence and consciousness. Of course, if we discover advanced life in another part of the galaxy, the aliens may simply supply us with all the answers to these vexing questions in one go!

Further Reading

Davies, Paul, *The Cosmic Blueprint: Order and Complexity at the Edge of Chaos* (London, 1987).

—— *The Mind of God: The Scientific Basis for a Rational World* (London, 1992).

—— *About Time: Einstein's Unfinished Revolution* (London, 1995).

—— *The Fifth Miracle: The Search for the Origin of Life* (London, 1998).

Richard Dawkins

RICHARD DAWKINS THROWS BACK HIS HEAD AND LAUGHS. HE'S IN THE MIDDLE OF RECOUNTING A CONVERSATION HE HAD RECENTLY WITH ONE OF HIS COLLEAGUES AT THE UNIVERSITY OF OXFORD. WHAT IF, SUGGESTED THE COLLEAGUE, DAWKINS WERE TO ANNOUNCE TO THE WORLD THAT HE HAD SUDDENLY AND MIRACULOUSLY BEEN CONVERTED TO CHRISTIANITY. INVITATIONS WOULD COME FLOODING IN—TO WRITE BOOKS, TO LECTURE ON THE AMERICAN CONFERENCE CIRCUIT, TO APPEAR ON TV CHAT SHOWS . . . DAWKINS COULD BECOME A MILLIONAIRE OVERNIGHT.

Of course, it's not going to happen. Dawkins, Oxford University's professor of the public understanding of science has no intention of ever turning to God. Indeed, he has been engaged in a running battle with the religious fraternity for years about where life comes from and how it developed. It's a row which has, at times, almost threatened to overshadow his work in biology, a field in which he has carved out a reputation as an influential thinker and popularizer of science, partly on the strength of a series of best-selling books.

Best-known of all of them, perhaps, is his 1976 work, *The Selfish Gene*, in which Dawkins advances the case for a selfish entity, the gene, that 'works' to preserve and propagate itself. The argument of the book is that the Darwinian theory of evolution via natural selection operates at the level of the gene and not at that of groups, species, or even individuals.

Dawkins defines natural selection as 'the non-random survival of randomly varying genetic instructions'. In other words, it is a two-stage process, the first being the production of random mutations in the genes of every new generation; the second the non-random effect of the environment on each individual gene, causing some to die, others to survive to pass on their mutations.

So life is simply a means of reproducing DNA. So lucidly does Dawkins explain this rather bleak message that *The Selfish Gene* has become a classic of science writing, the touchstone by which the stream of popular science writers now emerging judge their explanatory prose.

Many people want to debunk this theory, either because it contradicts their faith or through a vaguer feeling that it denies human dignity and freedom. Dawkins has devoted much time to vigorously repelling these attacks. He has explained, for example, how complex structures such as the human eye—structures that look as though they might have been designed by a higher intelligence—evolved by gradual, natural processes, the replication of a string of DNA molecules over generations.

Popular science writing has been one of the publishing success stories of the last decade. Yet, although thousands have bought and read books like *The Selfish Gene*, which set out scientific explanations for how we came to be, Dawkins believes that Darwinism has never been in as much need of advocacy as it is today. In saying this, he is thinking in particular of the power wielded by religious movements in the United States, where many children still learn that the earth and all its bounty were made by God in six days.

Paradoxically, whenever Dawkins takes part in American radio phone-ins he says he is encouraged by the audience's response. 'I find that some Americans hold the erroneous beliefs they do, largely because they have never been exposed to anything else. So when I try to explain something clearly . . . the reaction is not an attack but "Gee, that's right! Why didn't anyone tell me before?" '

Many scientists, though, are genuinely puzzled by the extent of Dawkins's embroilment in the row between science and religion. They worry that the broader scientific messages he sends out in his books suffer from the vehemence of his arguments with theologians.

Dawkins acknowledges that, to a certain extent, this is a problem. He remembers appearing on the BBC Radio Four programme *Desert Island Discs*. Presenter Sue Lawley spent most of the show focusing not on her guest's scientific views but on his attitude towards religion. Dawkins elaborates: 'You see, if you say something positive like "all living things are descended from a single common ancestor which lived about 4,000 million years ago and we are all cousins", well that is an exceedingly important and true thing to say, and that is what I want to say. Somebody who is religious sees that as threatening and so I am represented as attacking religion, and I am forced into responding. But you do not have to see my main purpose as attacking religion.'

There are aspects of religion that certainly annoy Dawkins—its use as a justification for wars and its interference with people's sex lives, for instance. He has condemned as 'disgusting' the Roman Catholic Church's stance against in vitro fertilization. But for Dawkins these irritations are 'separate from the more fundamental thing, which is a glorying, an exulting in the universe and a belief that the proper response to the universe is to try to understand it and not just let the sense of awe flatten one'.

Besides conducting a running argument with those who insist that God made the world and all its riches, Dawkins has also sparred with fellow biologists. *The Blind Watchmaker* includes an attack on the 'punctuated equilibrium' theory of American scientists Niles Eldredge and Stephen J. Gould. Eldredge and Gould believe that the evolution of a species may stand still for long periods and then change very fast in a short time—hence the phrase 'punctuated equilibrium'. Dawkins has no problem with this but feels that its revolutionary nature has been hugely exaggerated.

In his 1998 book *Unweaving the Rainbow* Dawkins launches a stronger attack on Gould's claims about the 'Cambrian explosion' and its alleged brief and exuberant burst of evolution.

At the heart of this attack is a genuine scientific puzzle. Precambrian rocks contain no fossils of many of the major *phyla*, or animal body-plans. Then, in the Cambrian period, fossils of the major animal body-plans appear, almost instantly in geological terms. Gould thinks the *phyla* appeared rapidly in the Cambrian era. Dawkins thinks they evolved more slowly during the Precambrian. Gould therefore has to explain a unique event in evolution, which never happened again. Dawkins has to explain why so few Precambrian fossils survive.

But there is a popular misconception of the 'Cambrian explosion' which says that each new *phylum* arose instantly, overnight, in a single generation. Dawkins regards this as dangerous anti-Darwinian nonsense. He could just about accept major *phyla* diverging from a single ancestry within 20 million years during the Cambrian period, to the point where a taxonomist would classify them as different *phyla*. But a mollusc does not give birth to a segmented worm.

The overnight-evolution theory has been promulgated by Gould's followers, and Dawkins holds Gould to blame. Yet Dawkins has seen many of his own ideas take wings and fly beyond his control, and he even has a theory to account for the phenomenon: he has coined the term 'meme', by analogy with 'gene', to describe a selfish, self-replicating idea, one that survives and evolves through generations.

Dawkins introduced the theory of memes, not as a theory of human culture, but 'to make the point that what matters in any theory of Darwinism is self-replicating information'. The human brain provides a new foundation for replication, not, this time, of genes—but of ideas. 'You have, in effect, a new primeval soup. Once you have got that new primeval soup, a new replicator could be the basis of a new Darwinism.'

Would Dawkins describe this reproduction of ideas as 'life'? 'It does not matter how you define life. If on another planet there is Darwinian replication and evolution, that is the interesting thing that I would like to find on another planet, and I think I would probably want to call it life. But if someone else preferred not to, that is their privilege.'

Long fascinated by computers, Dawkins has written computer

programs in an attempt to simulate aspects of evolution. His Blind Watchmaker program is a model of artificial selection with randomly generated variation and a limited kind of embryology that generates forms in two dimensions.

A proficient programmer, he started using computers in the age of paper tape: 'You would go to a computer room and you'd be smothered by paper tape and machines chuntering everywhere. You'd feel that you were in a room surrounded by all sorts of digital machinery—translators, copiers, and so on. And that is what a cell is like. A cell is a digital data-processing room filled with the equivalents of tapes and cards and bits and bytes floating around everywhere. At a poetic, metaphoric level, exposure to computers helps you to understand the world of DNA.'

Perhaps the one thing that unites Dawkins's critics and supporters is their respect for his writing skills. He himself is particularly proud of his writing in *The Blind Watchmaker* about bats and the way they steer and hunt by sound waves. Dawkins considers the bat's navigatory problem and the possible solutions, the problems those solutions introduce, and then the successive rounds of solutions which lead to exquisite adaptations in the bat's body over hundreds of years—adaptations that no organism could ever have achieved in a single leap.

The same approach is at work in the chapter on spider webs in Dawkins's fifth book, *Climbing Mount Improbable*. How would a silk-spinning predator make a living? Dawkins approaches the question from the point of view of the spider, living in 'a world of silken tension'. (The spider's sensory inputs, like the bat's, are wildly different from our own.) Every detail of web shape and construction is there for a reason, and each follows logically from what came before, yet not one is consciously designed.

Key Dawkins themes are here: the evolution of complex 'design-oid' structures by an accumulation of tiny modifications over centuries; the selfish genes which allow the male spider to be the female's lunch once the male has ensured that his genes will survive in her offspring; and the 'survival machines' that selfish genes construct for themselves. The web is as much a survival machine as the spider's body. The idea that genes reach out into the world and operate far beyond an organism's skin was the theme of *The Extended Phenotype*, his second and most technically academic book.

Dawkins has not carried out any laboratory research for several years now and, while such work still has an allure, he says he is too busy explaining and popularizing science to really miss it. 'On the whole I am happy doing what I am doing. If you do a good piece of research and you publish a very clever paper it is read by people and cited by them but the actual impact you have on most people's minds is not that great. But I think I have found my level and what I do best. I like to think that I may inspire other people to do the important work of research and maybe even encourage people to come into science.'

by Tony Durham and Kam Patel

A Riddle I Long to Answer . . .

RICHARD DAWKINS

I hope that during the twenty-first century the last vestiges of vitalism will be laid to rest. In practice this will mean that we finally understand what consciousness is and how it works.

Once upon a time people thought there was something special and unique about the chemistry of life. It was called organic chemistry for this reason. Now we understand that organic chemistry is just the chemistry of carbon. An organic substance is straightforwardly organic, whether or not it has any connection with life. Later, living matter was thought to be made of a special and unique material called protoplasm—quivering with throbbing vitality, and greater than the sum of its parts. Nobody speaks of protoplasm any more.

Living stuff is made of molecules like anything else, organized in a complicated way. The principle that drives the organization to become complicated is also now fully understood. This was the great contribution of nineteenth-century biology, in the shape of Charles Darwin. Twentieth-century biology, in the shape of James Watson, Francis Crick, and their colleagues, went on to remove the mysticism from the gene. Once again, any idea of a mysterious essence of life was replaced by something rigorously and totally understood, in this case the DNA code which is now about as mysterious as a computer tape.

Consciousness is vitalism's last desperate holdout. Twentieth-century biology still finds it genuinely mysterious, just as nineteenth-century biology found life mysterious. Life is no longer mysterious, and I have every hope that consciousness will go the same way. There are those who feel, along with Daniel Dennett in his book *Consciousness Explained*, that it is a non-problem. There are times when I think I see the force of this. At other times—when I experience the intense green of a banana frond in bright sunlight, or the smell of onions or the sound of bells, and when I reflect that all of these sensations are produced by trains of nerve impulses in the brain—at such times I see consciousness as a deep riddle that I long to answer, even while I find myself incapable of clearly formulating the nature of the question.

I believe that during the twenty-first century the ancient philosophical mind–body problem will be solved, and solved not by philosophers but by scientists.

Further Reading

Dawkins, Richard, *The Extended Phenotype* (Oxford, 1982).
—— *The Blind Watchmaker* (London, 1986).
—— *The Selfish Gene*, 2nd edn. (Oxford, 1989).
—— *River Out of Eden: a Darwinian view of life* (London, 1994).
—— *Climbing Mount Improbable* (London, 1996).
—— *Unweaving the Rainbow: Science, delusion and the appetite for wonder* (London, 1998).

Daniel Dennett

AS A SENIOR AT A TOP AMERICAN PREP SCHOOL IN NEW HAMPSHIRE, PHILOSOPHER DANIEL DENNETT PERFORMED THE BIT PART OF A BIBLE SELLER IN *INHERIT THE WIND*, A PLAY ABOUT THE TRIAL OF JOHN T. SCOPES.

Scopes was a schoolteacher thrown into jail for daring to teach evolutionary theory in an era when most Americans believed in the biblical account of Creation; God making the world in six days, including Adam, the first man. 'That is when I first became interested in evolution,' says Dennett, 'When I was learning about the background to the Scopes trial and hawking my bibles on the stage.'

Since then, he has become one of the most dedicated fans of the theory of evolution. 'To put it bluntly but fairly,' he writes in his book *Darwin's Dangerous Idea: Evolution and the Meanings of Life*, 'anyone today who doubts that the variety of life on this planet was produced by a process of evolution is ... inexcusably ignorant, in a world where three out of four people have learned to read and write.'

Dennett, 57, settles his imposing frame behind his desk at Tufts University, a private university outside Boston, where he has striven to be blunt but fair since 1971, when he first joined the philosophy department there. On his office wall are four reproductions of Marilyn Monroe—identical except that one is doodled with a moustache.

A pink plastic brain sits on the desk and behind it is a transparent model of a head, filled with wires. This last represents Dennett's controversial belief that what we think and remember and feel is no more than the sum of mechanical systems making up the brain.

So sure is he of this that, since 1993 he has acted as consultant to a team of professors and graduate students at the Massachusetts Institute of Technology, who are building a robot called Cog, a kind of artificial human. The plan is that Cog, despite being over six feet tall, will learn as a human infant does, acquiring new skills and knowledge as it interacts with its environment.

Cog's eyes are tiny video cameras, its ears are microphones aided by software able to discriminate speech sounds and its brain, the underlying operating system, is the size of a small cupboard. Work on Cog has been slower than expected and money to keep the research going has been hard to come by. Considering this, says Dennett, 'it is not surprising that Cog is still not ready for full-time life as an artificial infant.'

Dennett's international reputation is partly founded on his sympathy for scientific approaches towards problems originally defined by philosophers; problems such as 'what is human consciousness?' He argues that those studying the mind need make no distinction between what the philosopher David Chalmers has called 'Easy Problems', those concerning the mechanics of nerve and brain cells, and 'Hard Problems', which have to do with what philosophers call *qualia*—the way things look or smell or feel to us. 'Once all the Easy Problems are solved,' he argues, 'consciousness is explained.'

And the way the Easy Problems are solved is by applying the theory of evolution—which he defines as the process by which genes copy themselves to produce the genes of the next generation, making errors which lead to new characteristics and surviving, through natural selection, only if the errors are beneficial. Evolution, he says, is a 'universal acid' which eats away at some of the concepts of fields such as religion, ethics and economics, replacing them with a different— and unifying—explanation. No God or great designer is needed, simply a blind, mechanical algorithm.

Dennett is scathing about what he calls 'skyhook' explanations, such as religions, which rely on imaginary supports to describe why things are the way they are. Instead he favours concrete, mechanical explanations, which he calls 'cranes'.

'The traditional idea of a sacrosanct pearl of genius which is outside the realm of the mechanistic and is the source of creativity is just a hopeless idea, a fantasy,' he says. 'I know some people find this notion terribly offensive but that's too bad for them, since my job is to shock or cajole them out of their squeamishness.'

His is not a bleak view, he insists. He has no time for those who dismiss evolution as denying the existence of human creativity. 'Look how creative it is!' he cries. 'It has created every life form on the planet. If it can make a skylark, then it can make Keats' poem "Ode to a Nightingale". The skylark is at least as wonderful as the "Ode to a Nightingale" and the processes that have produced the skylark are, in the end, mechanical, algorithmic.'

Dennett's current research goes further, developing the idea that even free will is the product of evolution. It has evolved, he says, just as language has evolved and distinguishes human beings from animals in much the same way. He says that many people are terrified of accepting that free will is nothing more than the outcome of a mechanical process. It is this fear, in turn, that fuels their opposition to the idea that the brain is a computer and human consciousness a process that will one day be described down to the last detail by scientists.

'I want to say to them, you are absolutely right,' he says. 'Human freedom is the most important thing there is. But the way to protect it, to understand what it is, and preserve it, is not to try to dig a moat around it and protect it from science but to see how it evolves and see how it is, ultimately, perceived in a computational way.'

Dennett is not in the habit of telling his fellow academics they are right. He is a stalwart of academic argument, taking swipes at such revered thinkers as Stephen J. Gould, Noam Chomsky, John Searle, Roger Penrose. He tells a story about the immunologist Gerald Edelman being so angered by his criticisms that he once refused to shake his hand and walked out of the room when Dennett tried to introduce his wife to him at a meeting in Switzerland.

Darwin's Dangerous Idea was a sustained attack on people such as

Gould and Chomsky. Dennett says they misunderstood Darwinian evolution. In *Consciousness Explained*, published in 1991, he had already demolished the work of anyone who suggested that the phenomenon of human consciousness would never be totally explained.

What he most hates, he says, is an academic bully—'a silverback in his field that wilfully caricatures the views of the other side.' 'I get angry when people abuse their visibility and influence and authority,' he says. 'I really try not to do it myself. I have tried to make it a rule that I don't slam people unless they are really big and they really deserve it.'

On the other hand, he is quick to credit those academics who have helped and influenced his career—Gilbert Ryle, his doctoral supervisor at Oxford University in the 1960s, for example, or Richard Dawkins, the same university's current professor of the public understanding of science. He has relied on the goodwill of scientists to tell him about their field, he says, because his background was entirely in the arts. His father, who was killed in an aeroplane crash when Dennett was 5, was a professor of Islamic history. His mother was a librarian and textbook editor.

When the family returned to Massachusetts after living for a time in the Lebanon, Dennett and his sisters were sent off to Sunday School, where Dennett memorized the order of books in the Bible. One sister grew up to be a director of religious education. 'I understand the emotional treasures of religion as it were from the inside and so I am, I think, more sympathetic to religion than some of my colleagues and friends are,' he says.

This is curious in the light of past statements such as 'safety demands that religions be put in cages—when absolutely necessary.' He has often cited religion as one of the most dangerous 'skyhooks'.

But he says: 'It is certainly true that at its best religion can structure and fill human lives with meaning which enriches those lives and makes the world a better place and one tampers with that not just at one's peril but at the peril of everybody.'

The problem for an academic is that the comfort Dennett acknowledges that religion offers involves deliberate misrepresentation. But in day-to-day life white lies are sometimes necessary, he says. 'I think there is an understandable and somewhat defensible default position amongst academics and scientists that the truth is the highest good, and as long as it's true that's defensible. I don't believe that. I think that there are clear cases where we have to think much more deeply about the effect of truth on the world.'

This is because truths are so often misused and misunderstood. Dennett suggests: 'I think that academics should think about holding themselves responsible for the likely misinterpretations of what they say'—that they should present their truths to the world aware of how they could be distorted.

Dennett is, after all, a proponent of Richard Dawkins's concept of memes—ideas, fashions, behaviours, and skills passed on from person to person by imitation, competing against each other for survival and often being distorted in the process, subject to the evolutionary process in the same way as genes.

He is well aware of the dangers inherent in this. 'When you let fly a meme and watch it get mutated by the public you begin to see how something you have released into the cultural soup is taking on a life of its own, and perhaps developing in ways you don't like.'

Dennett's own acquisition of memes, through formal education, took place at a state school in Massachusetts, the New Hampshire prep school and then at the universities of Harvard and Oxford. A thoughtful child, who always wanted to be a teacher, he was also a tinkerer, making a device for producing pure oxygen, rigging up an alarm system to keep his sisters off his territory.

It was at Oxford, working for a doctorate in philosophy, that he developed a fascination with the mind and consciousness. There he began to devour medical textbooks and seek out people who could tell him about neuroscience. 'I decided that if I was going to do philosophy of mind, I had to know about relevant science,' he says.

At the same time, he did a bit of meme-spreading himself, teaching a course entitled 'ethics for printers' apprentices' at the local technical college where he discussed issues of abortion and euthanasia with young Oxfordshire men and women learning advanced techniques in linotype operation.

The upshot of his Oxford reading, followed by a period teaching at the brand new University of California, Irvine, was a series of books with enviable sales figures—*Content and Consciousness, Brainstorms, The Mind's I*, written with Douglas Hofstadter, and *Elbow Room*. By this time he was back home, teaching philosophy at Tufts.

He works, he says, 'in my own little personal imitation of natural selection'. 'I'm a pack rat. I scrounge around. I find lots of things that I think are interesting ... I pick up techniques and then of course visions, methods and then I chew on them and fiddle with them and let them spin around in my head and put them together in weird combinations.'

Dennett believes he hit on some really good ideas about evolution early on and has revelled in watching them develop over the years. While some theories need constant patching up and are ultimately doomed, he says, in others 'when you turn a crank, there are new insights, everywhere you turn you seem to be generating more good stuff.'

This is why he has never acted upon his vague fantasy of retiring to the family farm in Maine, where he spends his summers making cider and cutting up firewood. While he loves dealing with concrete issues in contrast to the 'preciousness' of academic life, he would miss the mental stimulation of his students too much.

So he keeps his back-to-nature instincts for the summer and meanwhile relaxes by whittling away at sculptures and small carvings. Recently, cruising with friends along the coast of Greenland, he encountered tupelaks—fetish figures, part animal, part person which Greenlanders carve out of reindeer antlers and often use as tools for cursing their foes. Fascinated, Dennett secured a supply of antlers and started working away at his own variations.

by Harriet Swain

Protecting Public Health

DANIEL DENNETT

As science and technology eliminate the barriers and friction that have heretofore constrained our human powers and thereby limited the scope of our moral choice, mankind's need for a reasoned, consensual, and open-minded ethics will become ever more pressing.

When we confront the fact that now we can choose many of the characteristics of our unborn children, can expose the contradictions and injustices in all our traditional religious views, can keep people alive for decades in conditions of slow deterioration—to take only three of the most difficult cases—we will have to decide what we ought to do, and this will require us to design and adopt a less tumultuous, more orderly process of determining our political will.

Being an optimist, I think we will succeed, but there is no denying the dangers. Fanaticism of every sort, on every issue, is bound to compete for our attention. H. L. Mencken once noted that for every difficult question there is a simple answer—and it's wrong. Unfortunately, many people cling to the simple wrong answers, and are even prepared to die—and kill—for them.

Creating and maintaining the democratic environment in which these momentous decisions are made will be one of the great challenges of the twenty-first century.

As communications technology makes it harder and harder for leaders to shield their people from outside information, and as the economic realities of the twenty-first century make it clearer and clearer that education is the most important investment any parent can make in a child, the floodgates will open all over the world, with tumultuous effects.

All the flotsam and jetsam of popular culture, all the trash and scum that accumulates in the corners of a free society, will inundate these relatively pristine regions along with the treasures of modern education, equal rights for women, better health care, workers' rights, democratic ideals, and openness to the cultures of others.

As the experience in the former Soviet Union shows only too clearly, the worst features of capitalism and high-tech wizardry are among the most robust replicators in this population explosion of memes, and there will be plenty of grounds for xenophobia, Luddism, and the tempting 'hygiene' of backwards-looking fundamentalism.

As Jared Diamond shows so eloquently, in *Guns, Germs and Steel*, it was European germs that brought Western hemisphere populations to the brink of extinction, since those people had had no history in which to develop tolerance for them. In the next century it will be our memes, both tonic and toxic, that will wreak havoc on the unprepared world. Our capacity to tolerate the toxic excesses of freedom cannot be assumed in others, or simply exported as one more commodity. The practically unlimited educability of any human being gives us hope of success, but designing and implementing the cultural inoculations necessary to fend off disaster, while respecting the rights of those in need of innoculation, will be an urgent task of great complexity, requiring not just better social science but also sensitivity, imagination, and courage. The field of 'public health', expanded to include cultural health, will be the greatest challenge of the next century.

Further Reading

Dennett, Daniel, *Darwin's Dangerous Idea: Evolution and the Meanings of Life* (London and New York, 1995).
—— *Brainchildren: Essays on Designing Minds* (Cambridge, Mass., and London, 1998).

Carl Djerassi

ARL DJERASSI CHANGED THE WORLD WHILE HE WAS STILL IN HIS TWENTIES. NOW 76, RICH AND RENOWNED, HE IS MAKING PROGRESS IN A NEW, LITERARY LIFE—WHILE STILL RETAINING HIS CHAIR IN CHEMISTRY, THE SUBJECT THAT BROUGHT HIM FAME AND FORTUNE.

Almost 60 years after arriving in the United States, Djerassi has no trace of an American accent, sounding like the Central European he is. He is one swirl in the vast stream of Jewish talent that fled Central Europe under the threat of Nazism. Britain, the United States, and other countries profited from this exodus, but in Djerassi's case the benefits have been global.

He worked in industry in the early years of his research career, with the US arm of Ciba, the Swiss drug firm now called Novartis, and has always been comfortable working in and with companies. In his late twenties, he took the apparently eccentric decision to quit the USA for Mexico to work at Syntex, a small pharmaceutical development firm that promised interesting work in a fascinating place. The result was the oral contraceptive, a technology known worldwide simply as 'the pill' and whose significance is still debated over forty years later.

In the late 1960s, doubts about the pill—an artificial version of the hormone that prevents women conceiving again while pregnant—were raised by a wide variety of groups. They ranged from Catholics, who object to interference with natural fertility processes, to feminists, who complained about a number of pill problems, not least that men had got together to produce a pill for women but not one for themselves. For the chemical contraceptive, particularly early versions of the pill, is not without its risks to women's health.

Djerassi points out that the current generation of feminists is kinder to his invention, seeing it as a device that gives them control over their fertility. His wife, Professor Diane Middlebrook, has encouraged his interest in feminist thinking. He teaches in the feminist studies programme at Stanford University in California, where he is professor of Chemistry, and is a strong supporter of promoting and recruiting women scientists. And he stresses that one thing alone that the pill has done—namely, reduce the number of abortions performed worldwide—ought to justify its invention in women's eyes.

Djerassi makes a prediction that in the twenty-first century, the pill will still be something taken by women, and that it will not be so very different from the one he and his colleagues developed in the 1950s. Though it may not, he acknowledges, be the most favoured form of contraception. For instance, there is promising research on new forms of female contraception, especially on the accurate detection of female fertility, which could make something like the rhythm method—a completely natural contraceptive technique—a lot more reliable. What is far less likely is a pill for men.

Djerassi says men's indifference to controlling their fertility is only part of the challenge. More important is their comparatively minor role in making babies: all men do is provide sperm. 'With women,' he says, 'there are far more possibilities because there are far more processes with which you can interfere.' Moreover, men are fertile for longer than women and might have to take the pill for 40 or more years. 'With a pill for women, all you are doing is preventing an existing egg developing. For a male pill to work you would have to shut down a major function, sperm production, for a long time'—which is potentially far more questionable.

But the killer blow is the lack of interest within the pharmaceutical industry in developing a male pill. 'Few big drug companies are working on fertility and even fewer on male fertility. The lead times are long, the risks are large, and the legal exposure is huge. Every male who took the pill and became impotent or got prostate cancer would get a lawyer the next day.'

Djerassi, who is examining male fertility, also downplays the possible effects of any male pill on population control. He says that there is little interest in controlling male or indeed female fertility among today's research funders. Neither the rich nor the poor worlds place contraception at the heart of their medical or population policies. Social policy, not pharmaceutical intervention, is regarded as the key to population control.

'Forget about the division between the "developed" and the "underdeveloped" world,' he says. 'It is better to think of the developed world—where the over-85s are the fastest-growing group—as "geriatric", and the underdeveloped world—where children are the fastest-growing group—as "paediatric".' In the geriatric world the emphasis is on trying to cure diseases of old age such as Alzheimer's, while in the paediatric world, eliminating malaria and other parasitic diseases ought to be the focus.

And even in those parts of the world where population is growing, the availability of contraception plays only a minor role in reducing the number of children families have. 'The fertility rates in Italy and Spain—the lowest in the world—show that cultural and economic factors are the key, that motivation, not the quality of contraceptives available, is what matters,' he says.

Djerassi's academic home since 1960 has been Stanford, an entrepreneurial university that is the research base for Silicon Valley. But his fortune was founded on the deal he made on going to Mexico. Increasing Syntex share prices brought in sums he is not prepared to talk about in detail but which must run into tens of millions of dollars.

At Stanford, and before, he published about 1,200 papers, and led major research groups. But in the interest of making the best use of his time—something of a Djerassi obsession—he has done little laboratory work of his own for decades. He does publish scientific papers, but his interests now centre on his life as 'a novelist who is starting to be a playwright and still does some chemistry'.

Despite this apparent change of gear, the work he is generating is in line with his earlier interests. His new métier is what he calls science-in-fiction, taking care to stress that this is not the same as science fiction. The aim is to create novels and plays that use real science to inform the public. Issues tackled so far include scientific competition, ageism and feminism in science, and the prospects of new reproductive technology. The last features in his penultimate novel, *Menachem's Seed*, and in his play, *An Immaculate Misconception*, which deals with ICSI, a technique for the artificial fertilization of eggs.

The books are a lively read, but feminists are unlikely to be impressed by some aspects of his work, such as the 'alpine breasts' of his heroine, who joins the music-lover's version of the Mile-high Club at a performance of the Vienna Opera. Last in the series of four novels is *NO*, named for the amazing physiological properties of nitric oxide, which is vital amongst other things for the male erection.

Human biology and fertility are a promising field for the novelist, Djerassi says, because of their social, political, medical, and religious ramifications. But he adds that they are often neglected because of their complexity. Today's writers have little interest in C. P. Snow-type bridging of apparently different cultures. '"Didactic" has turned into a dirty word for novelists,' he laments. But Djerassi has found fiction a useful device in teaching. His postdoctoral students write fiction to explore ethical problems in science.

As Djerassi sees it, novels are a way of expressing complexity, in contrast to science, where things tend to be either right or wrong. He says: 'The colour of most of these problems is grey, not black or white. There are political parties whose colours are brown, red, or green, but hardly ever a grey one. Perhaps we ought to adopt grey as the political colour of the twenty-first century.'

Djerassi regards himself as an active professor with no retirement date planned. But he has reached the point where he is thinking hard about what his life has meant, particularly since a 1985 brush with cancer. *Menachem's Seed* devotes a lot of space to Jewishness, which is becoming more important to Djerassi—in a completely secular way. He says: 'When I came to the US it was not something one advertised. I was the first Jewish chemistry professor at Wayne State (his previous institution) and at Stanford, but now there are many at Stanford, especially in the medical school.'

Despite his own literary endeavours, it is probable that Djerassi will be remembered by the artistic world as a patron rather than as a creator. He has long supported living artists. The suicide of his artist daughter, Pamela, in 1978, described movingly in his autobiography *The Pill, Pygmy Chimps and Degas' Horse*, spurred him to create a memorial to her. The result is the Djerassi Foundation, near San Francisco, on land bought with his Syntex profits. It runs the Djerassi Resident Artists Program for all kinds of artists from novelists to dancers, and over 1,000 have stayed so far. 'I don't believe in leaving much money to one's offspring. Instead I want to perpetuate something I feel strongly about, and as there is already lots of financial support for science, I decided to support the arts.'

by Martin Ince

The Century of A.R.T.

CARL DJERASSI

As a chemist involved in the first synthesis of a steroid oral contraceptive—a research achievement with major social implications during the past four decades—it is only natural to assume that I would wish to see fundamentally novel approaches to human birth control in the twenty-first century. But wishing will not make such approaches real, meaning applicable to millions of humans—as was, and is, the case with the pill.

The current lack of interest by much of the pharmaceutical industry in any advance in fertility control makes it very unlikely that fundamentally new approaches to human birth control (such as a contraceptive vaccine) will actually be made available to large segments of the human population. Nor is this situation likely to change, given the increasingly geriatric character of the Western world where most scientific research is performed.

By contrast, scientists and the public are extremely interested and involved in advances to improve, rather than to inhibit fertility, which leads me to a very different prediction for the first half of the next century, a prediction that I incorporated into one of the scenes from my 1999 play, *An Immaculate Misconception*.

MELANIE. I'm sure the day will come—maybe in another thirty years or even earlier—when sex and fertilization will be separate. Sex will be for love or lust—

FELIX. And reproduction under the microscope?

MELANIE. And why not?

FELIX. Reducing men to providers of a single sperm?

MELANIE. What's wrong with that . . . emphasizing quality rather than quantity? And as a bonus, burying once and for all that 'bashful egg/macho sperm' fantasy.

FELIX. Melanie Laidlaw's Brave New World! I wonder how many men will want to live there, . . . reproductively speaking.

MELANIE. We women will persuade them! I'm not talking of test tube babies or genetic manipulation. The woman will take a hormone cocktail to produce a dozen eggs or more, rather than the single one we release each month . . .

[Then] ICSI (intracytoplasmic sperm injection—the direct injection of a single sperm into an egg under the microscope) [will be carried out]. You'll inject each egg with a single sperm of her partner. I'm not promoting ovarian promiscuity, trying different men's sperm for each egg.

FELIX. 'Ovarian promiscuity!' That's a new one. (Chuckles). But what else is new?

MELANIE. Nothing is really new . . . it will just become routine. But each embryo will be screened genetically before the best one is transferred back into the woman's uterus. All we'll be doing is improving the odds over Nature's roll of the dice. That may be a new reproductive world, but brave? I'd call it intelligent. Before you know it, the twenty-first century will be called 'The Century of Art'

FELIX. Not science? Or technology?

MELANIE. The science of A . . . R . . . T . . . : assisted reproductive technologies. If my prediction is on target, contraception will become superfluous. Young men and women . . . and note my emphasis on young . . . will open reproductive bank accounts full of frozen sperm and eggs. And when they want a baby, they'll go to the bank to check out what they need.

FELIX. And once they have a such a bank account . . . get sterilized?

MELANIE. Exactly.

FELIX. I see. And the Pill will end up in a museum of 20th century ART?

MELANIE. Of course it won't happen overnight . . . . But A . . . R . . . T is pushing us that way . . . and I'm not saying it's all for the good. It will first happen among the most affluent people . . . and certainly not all over the world. At the outset, I suspect it will be right here . . . in the States . . . and especially in California.

FELIX. Predictions like yours and all that talk about reproductive empowerment make me wonder at times whether I'm too conservative to stay in this A . . . R . . . T business. Before you know it, single women may well be tempted to use ICSI to become a single mother. Who'd want that?.

MELANIE. . . . in other words . . . women like me.

It may seem ironic that a scientist, whose research contributed to preventing the creation of new life through ordinary sexual intercourse, now writes a play about the creation of new life in the absence of intercourse. But this play is meant to be serious and its predictions are indeed those of the author.

Further Reading

Djerassi, Carl, *Cantor's Dilemma* (New York, 1991).
—— *The Pill, Pygmy Chimps and Degas' Horse: an autobiography* (New York, 1993).
—— *The Bourbaki Gambit* (New York, 1996).
—— *Menachem's Seed* (New York, 1998).
—— *NO* (New York, 2000).

Andrea Dworkin

'PAUL GASCOIGNE SHOULD BE IN JAIL.' FOR AMERICAN RADICAL FEMINIST ANDREA DWORKIN, THE CASE OF THE GLASGOW RANGERS MIDFIELDER AND HIS ALLEGED BATTERING OF HIS WIFE IS SIMPLE. SHE DOES NOT BOTHER ABOUT QUESTIONS OF HIS SELECTION FOR THE ENGLAND FOOTBALL TEAM NOR THE REPORTED APOLOGIES FOR HIS BEHAVIOUR. 'IT'S ONLY WITH CRIMES AGAINST WOMEN THAT IT'S UP TO THE WOMAN TO COMPLAIN,' SHE EXPLAINS. 'THE STATE OPERATES IN THE INTEREST OF THE PEOPLE, AND IF IT IS STATE POLICY THAT MEN ARE NOT SUPPOSED TO BEAT WOMEN, THEN WHEN A WOMAN IS BEATEN THE STATE SHOULD MAKE THE ARREST AND THE ARREST SHOULD BE MANDATORY.'

The late 1990s saw several high-profile cases of wife battering. Dworkin was an obvious commentator. In the United States, she entered the O. J. Simpson debate, calling it an 'outrage' that Simpson's wife Nicole 'was beaten eight times before the police did anything to help her'. She deeply dislikes the glamour attached to depictions of brutality in both America and Britain, criticizing the 'endless entertainment' of violence in films.

Yet today she is, if not exactly on cloud nine, then at least mildly optimistic about the future. 'A lot of men feel almost dispossessed from the sense of masculinity that they thought they had a right to. Now they have an incredible opportunity to create equality with women,' she announces, over a glass of sparkling mineral water. 'I think that men should be thrilled that the kind of tyrant's role they assumed in the past has been taken away from them.'

Dworkin's positive attitude is surprising. Since the publication of her first book, *Woman Hating*, in 1974—a passionate exposé of violence against women—she has not been noted for her upbeat tone.

The American academic and commentator Camille Paglia, her most vicious critic, has accused her of 'wallowing in misery' and sharing with us 'her inability to cope with life,' while the popular perception in the media is of a large, ugly woman who 'hates men and sex'.

One reason for Dworkin's reputation as, according to Paglia, a 'clingy sob-sister' is the fact that her feminism owes its origins to personal experience. Although politically active from an early age and with a degree in literature and philosophy from Bennington College, Vermont, it was not until she was battered by her husband in her twenties, and was helped by a 'strange person who called herself a feminist', that she became interested in the movement. While other young women read the latest books of ideology which began emerging in the late 1960s, she was convinced only by practical detail. 'It was,' she explains, ' "you know he's going to hit you, he's going to kill you. I'm going to stop that from happening." And I had a lot of respect for that attitude.'

She went on to use her experience to help other women. 'I made a list of things I [had] figured [out] into [an explanation for] why, in this society, I might be in the situation I was in, and that list became the table of contents for *Woman Hating*.' Communication of pain and oppression was the *zeitgeist*. In 1976, two years after Dworkin's first book, the International Tribunal on Crimes Against Women was focused around what was termed 'personal testimony,' at which female victims of violence 'testified' about their experiences in a space free of men.

But Dworkin moved on from recounting her own experiences to collecting the stories of other women's oppression. In particular she became obsessed with what she saw as the violence of pornography. Her extensive research, which culminated in her seminal book *Pornography: Men Possessing Women* (1981), involved three years of looking at magazines and books, and venturing, armed with her notebook, into cinemas frequented by thrill-seeking men. The work gave her nightmares and isolated her from friends. 'I lived in a world of pictures,' she confesses, 'women's bodies displayed, women hunched and spread and hanged and pulled and tied and cut.'

Pornography, she concluded, was central to the sexual structures men had established. It confirmed the basis of those structures in dominance, force, even hatred of women. She resisted the notion of pornography as merely the representation of sexual acts. Her argument was not that pornography causes violence against women, but that it is violence against women. 'Pornography,' she explains, 'is documentary. It is not a matter of acting. It is not a matter of imitation. It is not a matter of modelling behaviour for other people. What is being done to the woman is really being done to the woman.'

Even women who are not depicted suffering torture but simply scantily clad are, for Dworkin, the victims of violence. 'The violence,' she asserts, 'isn't shown in the picture, but the violence is used to get the woman to make the pornography.'

So personal testimony was replaced in Dworkin's work by case history. *Pornography* is filled with the lurid detail of blue movie or magazine copy, aiming to turn 'the muck of real life' into 'something that we define and use rather than letting it define and use us'. Dworkin's rage led to practical steps to ban the offence. In 1983, she teamed up with Catharine MacKinnon, a law professor at the University of Michigan, to explore ways of legislating against pornography.

The result was what became known as the Minneapolis Ordinance, a draft bill defining pornography and allocating rights for damages to anyone either hurt in the making of pornography or assaulted as a result of pornography. The draft bill was rejected by government, but managed to show that a definition of pornography—and so legislation against it—was possible.

But Dworkin's rage, or her self-confessed 'misanthropy'—which resulted from her long and lonely study of pornography—also led her to the extreme, and many have considered, bizarre conclusion that nearly all heterosexual relationships are contaminated with the violence evident in pornography. 'Marriage as an institution developed from rape as a practice,' she asserted trenchantly in *Pornography*, and in her next book, *Intercourse*, she maintained that penetrative sex 'remains a means of physiologically making a woman inferior, communicating to her, cell by cell, her own inferior status.'

Camille Paglia's criticism of Dworkin's pessimism, though in many ways valid, carries little weight with Dworkin herself, who snorts in derision when she hears the name Paglia. Most feminists are, says Dworkin, agreed that Paglia's arguments are merely 'reiterations of standard anti-feminist rhetoric from an earlier backlash: that of 1890–1910'. But more damaging for Dworkin's reputation have been the criticisms and ideas of Naomi Wolf, author of the best-selling books *The Beauty Myth* and *Fire With Fire*.

Wolf protests that Dworkin's doom-laden descriptions of 'coercion, invasion, and one-sided objectification' in heterosexual relationships do not match her own experience. Dworkin, says Wolf, is 'obsolete' and does not represent the feelings of the majority of women. Dworkin herself says that most women are feminists but disguise their true feelings in mixed company. 'Most women,' she says, 'put on a good front, but when women talk among ourselves, the most man-hating women you will find are the most apparently normal, conforming women.'

Most telling of Wolf's criticisms is the description of Dworkin as a 'victim-feminist'. Dworkin's vision of 'overweening male oppression,' argues Wolf, means that she is unhappy about acknowledging that women can be anything other than powerless. Dworkin feels uneasy about acquiring power herself and treats unfairly those women who achieve success in what are regarded as conventional male spheres, says Wolf. She concentrates, in other words, on 'female victimization at the expense of female agency'.

There is some truth in this. Repeatedly, Dworkin dwells on her own personal struggle as a writer. An appendix to *Pornography* describes the emotional turmoil she went through in researching the book. And she says she has encountered many difficulties in finding a publisher for her work. She tells me that her book, *Life and Death*, published in the USA in 1998, has no publisher in Britain. It is all, it seems, a conspiracy of 'male figures at the top of the publishing hierarchy' who refuse the wishes of their more junior female editors. 'It's hard to publish books that in some sense bring people pain,' she says sorrowfully.

What annoys Dworkin about Wolf and others like her is that they speak for a particular social class for whom the acquisition of power is easy, and have no concept of the powerlessness of poverty. These women, mostly female academics and lawyers, have, she thinks, 'created some kind of protection for themselves and some kind of safety for themselves, and they don't even understand what they are talking about. They don't know what it's like to be poor; they don't know what it's like to be hungry.' Dworkin prefers to carry on collecting evidence from the street and from the brothel rather than to theorize and advise, shut away in libraries. And this is why she has a long-standing aversion to academics. 'When somebody publishes a book,' she says, 'I want to know who's going to live and who's going to die if we take that book seriously.'

Dworkin's sense of 'real life', then, is directly opposed to Wolf's, and it seems as if suddenly that vision is one that is becoming more widely shared. The organized rapes in Bosnia, and more recently it seems, in Kosovo, shocked women into realizing their continuing powerlessness after nearly three decades of the women's movement. Dworkin examines this in *Life and Death*. And in the present climate of morality and yearning for a return to a collective sense of right and wrong, Dworkin's black-and-white vision and what Paglia calls her 'Carry Nation' morality, which aims for state intervention to protect people, could gain a new currency.

There are even signs that Dworkin might be developing a more positive outlook. Her latest book explores the parallels between the pre-war treatment of Jews and the present treatment of women. The title *Scapegoat* ostensibly augurs a continuation of her litany of oppression. But the book examines the creation of the state of Israel and considers whether a similar institution for women, which avoids Israel's present faults, might be possible.

Dworkin muses: 'I'm interested in how the state of Israel changed the political situation of Jews all over the world and what a political solution for women might be. Because the land that we're fighting for is our own bodies. I don't think we want land. I don't think we want an army. We want this kind of sovereignty over our bodies that people essentially associate with state sovereignty. You know, boundaries that are respected, that you can't cross and if you cross them, you need the permission of somebody to cross them.' A sort of New Jerusalem for women: now that seems positive, perhaps even utopian.

by Jennifer Wallace

A New Jerusalem for Women

ANDREA DWORKIN

The changes that I hope to see are both simple and obvious. I want women to be literate, with shelter and food, independent of men, with a sense of the integrity of the female body, a new kind of sovereignty that makes the will of the woman primary: the internal assault of rape or the colonized body in prostitution would be stigmatized in international law, which would be driven by the experiences and needs of women.

I want the inner assault of rape to be widely understood as a form of lynching: with prostituted women as the bodies left hanging from the tree. I want to see a woman-dominated legal system in every country; and I want women to write the stories and the histories, to be the philosophers and theologians; I want the primacy of women in every culture and in every area of culture. I want women to have a sense of honour in relation to other women—and to children.

I want the sovereignty of a woman's body to be the primary sovereignty: not abrogated or compromised by the imperatives of male-centred legal systems. I want women to conquer the fear of male punishment; if that means the use of strategic violence, so be it.

I want countries in which women are sex-segregated, like Afghanistan and Saudi Arabia, to be pariah countries; I want the end of gender apartheid to be a demand of every political group and every country with any claim to a human-rights policy. I want infanticide of females to be stigmatized and illegal in the community of nations; I want the same for female genital mutilation. I want to see women freeing women from jails, purdah, sexual slavery, prostitution, and the domestic torture of marital battering and rape.

Patriarchy is dying a slow, slow death; but patriarchal power still tyrannizes women in households and in brothels. I expect to see deeper and more massive resistance from women in the next century, especially in the Third World.

Further Reading

Dworkin, Andrea, *Pornography: Men Possessing Women* (London, 1981).
—— *Ice and Fire* (London, 1986).
—— *Intercourse* (London, 1987).
—— *Mercy* (London, 1990).

Umberto Eco

THE ITALIAN PHILOSOPHER AND NOVELIST UMBERTO ECO GRANTS FEW INTERVIEWS THESE DAYS. TORMENTED BY OVER-ZEALOUS CRITICS, ACADEMICS, AND BUDDING WRITERS WHO PRESS HIM TO COMMENT ON HIS WORK, HE HAS WITHDRAWN INTO HIS SHELL, PREFERRING TO LEAVE THE INTERPRETATION OF HIS WRITINGS TO THE READER.

So I am delighted when, after almost four months of patient correspondence, Eco, professor of semiotics at the University of Bologna, finally agrees to meet me for an interview. The days before the meeting are fraught. I know that Eco has a weekly spot in the magazine *L'Espresso*, in which he comments on various aspects of Italian life, but, just before our meeting, a long article by him appears in *La Repubblica*, one of Italy's leading daily newspapers. It is about the way journalists misuse quotations. The fact that criticism of Italy's university system is also being widely aired in the media in the days leading up to our rendezvous does little to alleviate either his or my misgivings about our arrangement.

Yet, when we finally meet, at the university's Institute of Communication Disciplines, of which he is also the director, he is charmingly apologetic. '*Caro Pacitti*' (despite his excellent English he decides we should speak Italian), 'I should have to write a 700-page book to explain to you the various things that have happened to me recently, plus another just as long to answer the list of questions you have sent me.'

I reply that, given his publishing record, both would indubitably be best-sellers. He laughs, the ice is broken, and we are away.

Writing best-sellers has indeed become second nature to Eco. His three novels have been translated into more than thirty languages. The best-known, *The Name of the Rose*, a detective thriller set in the year 1327, tells how a medieval Holmes and Watson, sent to investigate a series of murders at a Benedictine monastery in northern Italy, are led to the culprit following their discovery that the pages of an ancient Aristotelian text have been treated with poison. Eco's book, which may also be read as a historical novel, a *Bildungsroman*, and a discourse on literature, invites the more philosophically inclined reader to consider the extent to which investigations into the true nature of objects in the world are obstructed by language itself.

His second novel, *Foucault's Pendulum*, centres on an over-erudite narrator, who leads the reader on a wild-goose chase of abstruse false tracks in an obsessive attempt to decipher a fragment of French text which he takes to be a dark conspiracy to rule the world but which, as his wife eventually informs him, turns out to be no more than a laundry list. This lesson in the foolishness of conspiracy theories and perverse readings of literary texts, culminates in the spectacular death of the hero's friend, strangled by the actual Foucault pendulum in Paris.

And in his most recent novel, *The Island of the Day Before*, a lyrical tale clearly inspired by Edgar Allan Poe's *MS Found in a Bottle*, a Piedmontese shipwreck survivor in search of love and the meaning of life, finds himself on board a deserted ship in the South Pacific in 1643, headed for a mysterious island of salvation. Love letters and memories of unquenched passions alternate with images of the clocks and maps used by an old Jesuit to sound the depths of the universe as an obsessive narrator on an impossible voyage struggles to put names to the stream of unknown objects he encounters.

Despite the fact that his novels have sold millions, Eco seems untouched by the fame they have brought, preferring to speak simply of his 'narrating' as a separate activity, secondary to his academic study of philosophy. At the moment, he reveals, there are no more novels in the pipeline; his plans are to concentrate on philosophy.

Nonetheless, he admits that his novels can be read as a prelude to his academic writing. They too are about the search for meaning in life and elaborate on what are for Eco recurrent themes—the complex relationship between appearance and reality, between interpretation and misinterpretation and, between words and the objects to which they refer.

Umberto Eco was born in Alessandria in Piedmont, forty miles east of Turin, in 1932. He attended the University of Turin where he studied under the celebrated Italian existentialist philosopher Nicola Abbagnano, taking his degree in philosophy in 1954 with a graduation thesis on 'The Aesthetic Problem in Saint Thomas Aquinas', which marked the beginning of a lifelong interest in medieval history and thought.

For a few years after leaving university Eco worked as cultural editor for the Italian national radio and television network, RAI, leaving, in 1959, to become senior non-fiction editor for Bompiani publishers in Milan. He was a founder member of 'Gruppo 63', a radical intellectual group of the 1960s.

In 1961 he became lecturer in aesthetics at the University of Turin while still working for Bompiani. He subsequently taught aesthetics at Milan Polytechnic, visual communication at the University of Florence and semiotics at Milan Polytechnic. In 1975 he was appointed professor of semiotics at the University of Bologna, the position he currently holds.

Semiotics is commonly defined as the theory of sign systems in language. Eco, however, author of *A Theory of Semiotics*, is keen to extend this definition. 'Semiotics attempts to explore all possible systems of communication,' he says. 'This leads to a philosophy of language which goes beyond verbal structures and usage. It includes all forms of language, from gesture to the visual image.' He is unperturbed by the snobbery many philosophers show towards the study of semiotics, with some even arguing that its investigation cannot properly be considered philosophy. 'I believe that semiotics is the only form of philosophy possible today,' he says firmly.

For Eco the central issues of philosophy are: how do we assign meaning and how do we perceive things? One of his most recent books, *Kant and the Platypus*, a collection of six philosophical essays, addresses the problem. In it Eco elaborates on the question of how the German philosopher Immanuel Kant would have set about ascribing identity to a platypus, an animal he would not have been familiar with, and which would have seemed to Kant to be nothing more than an assemblage of parts from other creatures.

In *Kant and the Platypus* Eco concludes that words acquire meaning as a result of negotiation within a particular community of speakers. Objects and speakers do not come with God-given identity tags; rather identity is conferred by arrangement and consensus. Yet, argues Eco, amongst the many possibilities opened up by such a process of negotiation, there are basic facts that place limits on what words can be said to mean.

Unlike his previous highly technical philosophical writing, *Kant and the Platypus* bristles with philosophical parables. Eco says that the new formula was a deliberate move away from his more difficult writing. 'I wanted to write a book which differed from my previous ones in being non-systematic rather than systematic,' he says. 'These parables and anecdotes are useful for the formulation of questions and problems and go some way towards the avoidance of technical terminology.'

In one example we are asked to imagine Ptolemy, Galileo, Isaac Newton, Kepler, and Epicurus together at the top of Mount Arcetri near Florence. All of them are looking at the sun, though each one espouses a different cosmological theory. 'While the contractual aspect of each one's system of ideas determines his mode of perception,' Eco argues, 'what cannot be denied is that they are all thinking about that red object out there, after which negotiation [about the nature of the sun's identity] will obviously begin.'

He goes on: 'I believe that much of the contemporary philosophy of language has rather forgotten that, in the end, they are all talking about the same thing. Had Ptolemy pointed instead at a silver circle the others would have said, "No, that's the moon, we're talking about that object over there." I am especially interested in those things which cannot be said [about the sun or about any other empirical object] because they are not true to the facts.'

Fanciful readings of works of literature have provided Eco with plenty of examples of 'things which cannot be said'. As he points out, Dante's *Divina Commedia* has been plundered by Freemasons and Rosicrucians who have written elaborate justifications for their convictions that esoteric symbols are discernible in the text. 'Their readings are examples of things which cannot be said because they go beyond the limits,' says Eco. 'The text simply does not say what these readers suggest. There are certain lines of resistance in a text just as there are in reality.' Eco is suddenly emphatic, agitated by his well-known intolerance of perverse interpretations of literature.

Lowering his voice again he adds: 'In my book *The Limits of Interpretation*, an attack on some forms of literary criticism, I argued that every text is open to an infinite number of possible interpretations, yet there are certain limits.' And the limits are imposed by empirical facts; an interpretation that does not fit the facts of a piece of writing cannot be accepted. '*Kant and the Platypus* is the transposition of this problem from texts to reality,' he adds.

Eco accepts that the need to draw boundaries is a constant in both his literary and philosophical thinking. 'I find I am discovering more and more,' he says, 'that the basic tool in philosophical thinking is common sense. For the past thirty or forty years French philosophy has forgotten common sense. I think it's high time that common sense, so fundamental to the history of philosophy, was reintroduced to the scene. Aristotle was, above all, a man of extraordinary common sense, as was Aquinas.'

Rejecting the dictum that philosophy begins with the loss of common sense, with the need to question the accepted, he says: 'Philosophy goes beyond common sense in that philosophers question facts or ideas that others take for granted. But to go beyond does not necessarily mean to reject. Nor does it mean to go against. What it does mean is that the philosopher continues to use common sense in order to tackle problems that everyday life does not raise.'

So does Eco's view of philosophy allow for the possibility that 'truth' exists? Truth is a concept scorned by postmodernist thinkers, who hold, rather, that there are many truths, each dependent on an individual's viewpoint and all constructed by consensus within a community. 'I hold that there can be no truth which is not the result of people interpreting reality, and hence resulting from a social contract,' says Eco. 'But when we come across those lines of resistance that prevent us from making certain statements, that is the closest we can get to truth. There is something in reality that decrees: "No, you cannot say this." Negation is the closest thing to truth. What is true is that you cannot say something because it crosses the limits.'

On the subject of the many arguments and feuds that have riven philosophy in recent years Eco is phlegmatic. He is in favour of international collaboration and says that when philosophers meet they often find that they agree on many issues. 'In 1995 three-quarters of Britain's Cambridge University philosophy faculty opposed the award of an honorary degree to the French philosopher Jacques Derrida. Academic compartmentalism leads to a kind of defensiveness. But when we meet we find we have much in common.'

He has even tried to promote international collaboration by founding a university in the tiny independent state of San Marino ten years ago. 'We felt we could operate more efficiently by freeing ourselves from the bureaucratic shackles of the Italian state. For a while it worked well for us—the British historian Eric Hobsbawm and the American philosopher Hilary Putnam were among the many international figures who attended our conferences. But differences arose over the direction we should follow, over future programmes. It all became too much for me and I withdrew under the pressure.'

As the University of San Marino continues to battle on without him, Eco has his work cut out at Bologna, where thousands of students compete every year for just three hundred places on his course—the most popular in Italy.

It is a mark of his fame that my last question proves the most trying, throwing Eco into visible disarray. How many honorary degrees does he now hold? Summoning his secretary to bring him his website address he suddenly remembers that it has not been updated for months. 'Twenty-two at the last count, including three from Britain.' The other nineteen span twelve countries, 'but I have a feeling that there may be a couple more not yet on the list. Can I e-mail you an attached file?'

by Domenico Pacitti

Never Fall in Love with* Your Own Airship

UMBERTO ECO

At present, many scholars are working to understand how the human brain functions, including the question of how language is produced. In particular, they are trying to understand the mystery of human consciousness.

In my book *Kant and the Platypus*, I refer to people working on the secrets of the human brain, or as I describe it, 'the black box'. The scientists engaged with this question are proceeding in leaps and bounds.

I do not work directly on unlocking the processes of the black box. Unlike cognitive scientists I do not study synapses and neurons. Rather, I study the output of and input to the black box.

But black box research is progressing in such a way that it is gradually changing many of our ideas. It is difficult to say today in what precise ways these discoveries in the cognitive sciences will change ideas in philosophy, semiotics and linguistics. Perhaps completely new facts and insights will emerge; some of the divisions of knowledge may disappear.

I must say, I try not to make these predictions. Just imagine what it was like when the airship was invented. What a wonderful thing, people thought, to be able to travel through the air just like a bird. And then it was discovered that the airship was a dead-end invention. The invention that survived was the aeroplane.

When the first airships appeared, people thought there would subsequently be a linear progression, an advancement to more refined,

* Transcript of part of an interview with Domenico Pacitti

swifter models. But this did not happen. Instead, at a certain point there was a lateral development. After the Hindenburg went up in flames in 1937, [killing 35 people], things began to move in a different direction. At one time it seemed most logical that you had to be lighter than air in order to fly in the sky—but then it turned out that you had to be heavier than air to fly more efficiently.

The moral of the story is that in both philosophy and the sciences you must be very careful not to fall in love with your own airship.

Further Reading

Eco, Umberto, *A Theory of Semiotics* (Bloomington, 1976).
—— *The Name of the Rose* (London, 1983).
—— *Foucault's Pendulum* (London, 1989).
—— *The Limits of Interpretation* (Indiana, 1990).
—— *The Island of the Day Before* (London, 1995).
—— *Kant and the Platypus: Essays*, trans. Alastair McEwen (London, 1999).

Francis Fukuyama

L IFE IS GOOD FOR FRANCIS FUKUYAMA. HE LOVES HIS JOB AS PROFESSOR OF PUBLIC POLICY AT THE INSTITUTE OF PUBLIC POLICY, GEORGE MASON UNIVERSITY. HE LOVES WRITING BOOKS ABOUT HOW WE LIVE, WHICH PROMPT COLUMNS OF NEWSPRINT WHEN THEY ARE PUBLISHED. HE LOVES SPENDING SPARE TIME WITH HIS THREE CHILDREN, WHO APPEAR IN FULL COLOUR ON HIS WEBSITE WITH ACCOUNTS OF THEIR INTERESTS AND ACHIEVEMENTS.

Best of all, he is lucky enough to be writing in a place that, politically at least, just happens to have reached a peak. For him, liberal democracy of the kind that exists in the United States is 'the only viable alternative for technologically advanced societies'. In other words, as he will long be remembered for saying, liberal democracy represents 'the end of History'. And for a time, at least, as the formerly Communist regimes of the Eastern bloc collapsed like a row of dominoes, he appeared to be right.

Now, in his latest book, *The Great Disruption: Human Nature and the Reconstitution of Social Order* Fukuyama claims things are looking up—at least in the United States—in the moral sphere too. While the march of political development is progressive, the development of morality is cyclical, he argues. But, after 'the great social disruption' of the 1960s, caused by technological advances such as the birth control pill, Americans at last seem to be on the upward curve.

Rates of increase in crime, divorce, and illegitimacy have slowed, numbers on welfare rolls have dropped and men and women both seem to be taking more responsibility for their children. This is because, he explains, human beings are, by nature, social creatures, whose 'most basic drives and instincts lead them to create moral rules that bind them together into communities'. Picking up on the theories of the biologists, he argues that it is in our genes 'to invent moral rules to constrain individual choice'.

The only cloud on Fukuyama's rosy horizon is that nobody likes an optimist. Once described as someone who made *Candide*'s Dr Pangloss look whingeing, he has been accused of telling the American elite what it wants to hear, of boosting the confidence of Western capitalist nations with his reactionary views about the nuclear family, and his insistence that capitalism is not only successful but right.

The criticism doesn't seem to bother Fukuyama in the slightest, who says he is simply being realistic. 'I don't start out wanting to be optimistic,' he says. 'It is not easy being an optimist. Mankind has come through a really horrible century—at least the first two-thirds of it. It is much easier to be a pessimist who is proved wrong, because then at least you achieve a certain sense of moral gravity. But realistically, if you look at things at the end of the twentieth century, politically, economically, and socially, they have worked out much better than anyone would have thought.'

This goes even for the poorer sections of society. 'The shift towards freer markets and more open, democratic political forms has been broadly empowering for many people, and not just for the crowd that stands at the top of the social hierarchy. If you look at the economic development of East Asia you look at millions and millions of people who were living in poverty who are now leading middle-class lives and have done well.'

Fukuyama's willingness to express the unfashionable view could suggest a low *thymos*, as he calls it—the desire for recognition that drives human beings on. But it is this very knack for controversy which has brought him worldwide recognition over the last ten years.

He has never quite looked back from the publication of *The End of History and the Last Man*, which started life as a lecture to the University of Chicago in 1989. 'I was asked to lecture in a series on the decline of the West,' he says. 'I said I would give a lecture but that it would not be about the decline of the West, it would be about the victory of the West.'

In it, he talked about the triumph of liberal democracy, which had settled all the big questions of History with a capital H. Taking Hegel's belief in the idea of History as a single, evolutionary process, he argued that this evolution had now taken place. Liberal democracy was the best political system for advanced technological societies and capitalism, which was the most efficient way of exploiting technology, was the best economic system.

The stir caused by his lecture became a sizeable whirlpool following the fall of the Berlin Wall a few months after he delivered it. Several ideological wars since, including the Balkans, have been cited in an attempt to prove him wrong. But he sticks to his principles. Recent wars in Kosovo, Rwanda, Liberia are the skirmishes that happen on the periphery of modern states and in states that are not yet 'at the end of History'. US president Bill Clinton and Britain's prime minister Tony Blair were wrong to suggest that the Balkan conflict could be a tinderbox drawing the industrial powers into a major contest.

In his second book, *Trust: The Social Virtues and the Creation of Prosperity*, Fukuyama moved down a scale to look at the 'best' cultural supports for the capitalist democracies at History's end. For him, 'a nation's well-being, as well as its ability to compete, is conditioned by . . . the level of trust inherent in the society.' Societies that have strong families but relatively weak bonds of trust among people who are not related tend, he argues, to be dominated by small, family-owned and managed businesses. Prosperity is more likely in countries where people are used to associating in groups larger than the family. So Japan and Germany can support big firms such as Sony and Siemens because their societies possess a strong culture of trust and mutual support. France and Italy have done less well economically because they have less of a tradition of civil society.

The Great Disruption moves a step down again, to look at the kind of domestic arrangements which best support stable societies. To do this, he spends some time exploring 'human nature', taking a close look at the biological forces that he believes determine family life.

He is reluctant, because of the fierce reactions it provokes, to talk about the biological origins for male/female characteristics. Instead, he suggests it may be better to talk about female 'socialization'. Yet his book shows no such qualms. 'While the role of mother can be safely said to be grounded in biology, the role of [late twentieth century] father is, to a much greater degree, socially constructed,' he writes. Men 'have a biological disposition to be more promiscuous and less discriminating than women in their search for sexual gratification.'

Even more controversially, he blames recent breakdowns in family life on the impact of the pill and the sexual revolution, and cites evidence linking higher female earnings to rising divorce rates and a trend towards child-bearing outside marriage.

His interest in the biological sciences is relatively recent. Introduced to some of the literature by friends, he is now a strong advocate of the importance of biological insights for social scientists. 'The more I read the more it became evident that the social sciences were operating on a principle that was ideologically based. This was a reaction to the Holocaust and to the genuine misuses of biology by assorted racists and bigots. Social science had been turned in the opposite direction—to say that one's genetic basis meant nothing,' he says. 'But there has been a genuine revolution in biological discoveries. The social sciences have to adjust to this.'

He concedes that it is a sensitive area and that the 'bell curves' he favours, showing differences between male and female reproductive strategies, for instance, apply statistically only to large populations so, in a sense, are irrelevant for individuals. 'But I think this does have implications for social policy because social policy deals with populations,' he says.

Biology can help in thinking through ethical and political problems because 'it shows you that there are certain constraints in social engineering that limit the kind of society you can create'. For example, socialism began with the mistaken premise that human beings were altruistic or could be made to be more altruistic than in fact they were. Nature therefore 'provides a negative lesson that there are certain types of utopias that aren't realizable'.

Fukuyama's enthusiasm for biology shows something of the convert's zeal. He comes from a long line of academics. His maternal grandfather was a prominent economist in Japan, who passed to Fukuyama a first edition of Karl Marx's *Das Kapital*, bought on a study visit to Berlin before World War I.

His father, a second-generation Japanese American, was a congregationalist minister and academic sociologist. 'There is a certain sort of Protestant who is very moralistic about a lot of things and he was of that category,' says Fukuyama. 'He was way to the left, as it turned out, of where I am so we had a lot of political arguments.' His mother, who had an MA in social work, worked 'intermittently' with his father.

He took Classics as an undergraduate at Cornell, where he studied political theory under Allan Bloom, before making what he called 'a wrong turn' by studying comparative literature at Yale. This involved a period in France being taught by Jacques Derrida and Roland Barthes, all of which he thought was 'totally bankrupt' intellectually, so he quit to study international relations at Harvard.

This led to a job working for the State Department, before his 'End of History' lecture catapulted him into academic life without all the tedious specialization which he complains is usually necessary for tenure in America. 'The most interesting things are interdisciplinary these days,' he says, although this attitude has not made him popular with academics.

He does not miss one bit the practical side of politics. After all, it is an exciting time to be thinking about politics, what with the impact of globalization and the fallout from feminism. Both phenomena, he says, are likely to lead to fewer wars. Commercial competition is likely to take over from warfare as a mechanism for new technological developments. And it is well known that those responsible for war are predominantly young men, while the populations of the developed world will be proportionately older rather than younger in the future as people live longer and birth rates fall.

Terrible things could happen, he admits. There could be a nuclear explosion. Genetic engineering could wipe out human life as we know it. But there will remain 'very powerful innate human capacities for reconstituting social order'.

Personally, he remains content with teaching and writing and fiddling about with his computer, making ever fancier virtual furniture. He used to make real reproductions from wood, after coveting the Sheraton and Hepplewhite antiques in the State Department. Naturally, his new, computerized, hobby does not let him keep the furniture. But this way, he says, he manages to avoid splinters.

by Harriet Swain

The Politics of Women[*]

FRANCIS FUKUYAMA

Politically, the struggle has shifted from the old Cold War divisions to a struggle over globalization. In a way it is a working out of my *End of History* hypothesis: all we have now is the global economy that defines our way of life and is reshaping politics and economics around the world. In spite of all the turmoil that took place in the financial markets in 1997 and 1998, this is still on track. Not only that, but it is all but inevitable that global markets will be further integrated and that there will be more competition. This is driven by the evolution in information technology that is democratizing access to information, but that also knits people together in ways they sometimes do not like. It is very hard to see how that is going to be undone. And the further integration of global markets will further enforce the norms and institutions of the liberal democratic West.

The real question in the future will concern possible backlashes against this kind of globalization: whether it will lead to countries trying to exit the system entirely—which I think is something very difficult to do—or whether it will lead to reactions like those of Ramzi Yousef who tried to blow up the World Trade Center Tower out of resentment against the United States.

What is driving this whole process is people's desire to live in a modern world, so they have access to things like McDonalds and consumer goods, and so they can participate in their political system. If people suffer corrupt governments globalization gives them alternatives, such as the ability to bring in international human rights monitors. It is not something that is being imposed on the rest of the world by the West, but is truly driven by people voting with their feet in less

[*]In association with Harriet Swain.

developed countries. I think certain states are going to be spoilers [in this trend]—China is probably the most significant because it is the most powerful—and perhaps Russia will be too.

Thailand, Indonesia, and South Korea went into a very severe currency crisis at the end of 1997, but Thailand and South Korea, for all practical purposes, have now recovered and, if anything, are better off than before. Their situation was so serious that it forced them into a series of political reforms. Thailand now has a constitution that is the most democratic in its history. South Korea has undertaken some major structural reforms in liberalizing its economy. In general, it was Asian authoritarian governments that posed the greatest challenge to Western liberal democracy because countries with such governments seemed to be able to master modern science and technology and do so in a culturally different way. But in fact, the Asian crisis shows that in the long run these countries cannot maintain a culturally distinct set of institutions. Everyone is being forced to adopt more Western standards of transparency and openness.

I believe that while political and economic history tends to be progressive and we are moving towards more democracy and a common framework of institutions for all societies, moral values are cyclical. It is my belief and hope that we hit the bottom of the current cycle sometime in the early 1990s and that countries that experienced the 'Great [moral] Disruption' of the 1960s are re-forming themselves so that they can adjust to the relocations that were caused from shifting out of the industrial era.

Culturally, I think there are a number of trends based on demographics that will, particularly for Europe, have fairly serious consequences in the future. One is the result of falling fertility. In Italy, Germany, and Japan the drop in fertility has been so dramatic that these countries are going to be losing well over 1 per cent of the population per year, every year, as they go into the twenty-first century. We already know that they face a very severe social security liability as a result of this. The only way they can make up for falling population is by importing workers, unless there is a dramatic increase in fertility which I do not think anyone expects to happen.

In this kind of world, the English-speaking countries will do extremely well because they have a long history of accepting immigrants and assimilating them into a broader mainstream culture. But countries like Germany and Japan, that still understand citizenship in ethnic and racial terms, are going to have difficulties. Even France, which is somewhere between the English-speaking world and the rest of the continent and has been relatively liberal in the past, still has a big problem with the *Front National*, Jean Marie Le Pen, and the backlash that Third World immigration has created. Unfortunately, because France is also pursuing such thick-headed economic policies that keep its unemployment rate high, this produces a very flammable mixture.

But it is hard to see how Europe can avoid becoming much more culturally diverse in the next couple of generations because the native-born populations are going to be shrinking at such a rapid rate.

Properly managed diversity can be extremely valuable. If you look at Silicon Valley in California, something like close to half the engineers there were born outside the United States. Economically, such diversity has a tremendously stimulating effect. Culturally, it is harder to say what will happen because some cultures assimilate much more readily than others. To this day, the black population in the United States, for all of the effort and emotion that has been put into achieving social equality, still remains very troubled, whereas Asians, for the most part within a generation or two, intermarry and get absorbed into the larger culture. The Hispanic and Asian immigration in the United States has tended to strengthen more traditional family values because in Hispanic and Asian cultures such values have not disappeared.

But we also face other types of cultural change. In some ways, you cannot have an *end of history* unless you have an *end of science* and we are just on the verge of another big explosion in scientific knowledge that comes out of the life sciences—the potential for bio-technology to change some very basic aspects of human existence. It could lead to an end of human beings as we understand them. We now have the possibility of altering the germ line, which means it is possible not only to alter the behaviour and characteristics of the present gener-ation, but to create inheritable characteristics. This in effect means changing human nature. This may make possible certain kinds of social engineering that heretofore have failed.

The other great change is continued feminization of political life, both at a domestic and international level. There is a lot of evidence that women approach politics, particularly international politics, very differently from the way men do. In the United States, women have consistently failed to support military intervention, defence, and power competition generally in the international realm, by about 9–10 percentage points for as far back as this sort of polling has been done.

Moreover, with the overall demographic shift, populations are age-ing. In another fifty years, countries in the developed world will have median ages somewhere in the fifties. The combination of the greater impact of women on the political system and the shifting of the median age upwards is going to lead to a very different kind of politics. Let's face it, most of the trouble in the world is caused by young men, or else Saddam Hussein types who want to lead young men into various kinds of adventures.

Certainly, societies in Europe, North America and North-East Asia, are moving very rapidly away from politics involving military intervention, defence, and power competition. I think that is good for democracy, because one of the things that democracy implies is fully equal female participation in the political process. But I think the very nature of the politics will change as a result.

Further Reading

Fukuyama, Francis, *The End of History and the Last Man* (New York, 1992).

—— *Trust: The Social Virtues and the Creation of Prosperity* (New York, 1995).

—— *The Great Disruption: Human Nature and the Reconstitution of Social Order* (London, 1999).

J. K. Galbraith

MOST ECONOMISTS, WHATEVER THEIR POLITICAL PERSUASION, COULD ONLY DREAM OF BEING JOHN KENNETH GALBRAITH. THE SCION OF SCOTTISH SETTLERS IN SOUTHERN ONTARIO, HE SHRUGGED OFF THE DRY FORMULAE AND FIGURES OF HIS CALLING TO BECOME AN AMERICAN PRESENCE, WHO TOURED INDIA WITH JACKIE ONASSIS. HE WAS AN ADVISER TO PRESIDENTS AND BEST-SELLING AUTHOR. IN LETTERS FROM J.K.G. TO J.F.K., HE REVELS IN HIS OWN EGO. 'GALBRAITH'S FIRST LAW,' SAYS AN EMBROIDERED PILLOW, LODGED ON A BOOKSHELF IN HIS CAMBRIDGE, MASSACHUSETTS SITTING ROOM: 'MODESTY IS A VASTLY OVERRATED VIRTUE'.

Forty years after the publication of his hugely influential book, *The Affluent Society*, and nearly twenty years after he completed his memoirs, the press still come calling for J.K.G.'s wit and vim. Curiously, it is often the television teams, whose vapid coverage he has long affected to despise. Most recently, the 90-year-old economist and author of *The Great Crash*, among a slew of other books, was fielding enquiries in the fallout from the slump in the Asian stock market.

For the record, Galbraith is extremely reluctant to make any prediction for the economic future—because only the wrong ones are remembered, he says. He contents himself with warning that the United States is undergoing 'extreme stock market speculation', with far more mutual funds (whereby investors buy shares in a company which in turn invests their money in a spread of stocks and bonds) than there is 'intelligence to manage them'. He adds: 'Cycles of euphoria and recession are a feature of capitalism, and have been for hundreds of years. We should be fully aware of the dangers of economic difficulties that will not be confined to Malaysia.'

In 1958 the publication of *The Affluent Society* secured Galbraith's place as one of the titans of American economics, albeit from an unapologetically liberal perspective. On its publication, he recalls in his memoirs, it was received by *Time* magazine as 'a well-written but vague essay with the air of worried dinner-table conversation'. Other reviews were ecstatic.

Galbraith still regards the general theme of *The Affluent Society* as the key legacy of his work: 'that in a modern country, particularly the US, we have a higher standard of private consumption than we do of public goods,' he says. 'We have clean houses and dirty streets. We have expensive television and poor schools. Recreation and education, particularly, on which the poor depend, are much more meagre than that which is available in the private sector to the rich.'

The disparity reflects 'the inescapable tendency of countries, as they grow rich, to become more negligent of the poor,' he continues. 'The poor become a voting minority, and the rich and the affluent come increasingly to attribute their good fortune to their superior character or intelligence ... the consequence of that is the terrible situation in our cities, the disgrace to our country.'

Yet, despite his international reputation, J. K. Galbraith, it appears, is no longer required reading in the mass market. My local bookstore in Los Angeles carries neither *The Affluent Society* nor *The New Industrial State*. Enquiries farther afield elicit a similar response. The bookstore staff are faintly embarrassed by their absence, but the truth is that Galbraith's take on economics and society is no longer terribly fashionable.

As he entered his retirement, as an 'abiding liberal', the era of Big Government, of sanctioned state spending on public services, was declared over, first by Republican president Ronald Reagan, then by the Democrats. 'Liberal', in America, became a dirty word for Republicans and an embarrassment to centrist Democrats. The Democrats' fear of being seen as 'soft', a fear Galbraith denounced in military and foreign policy in the 1960s, now appears to extend to social and economic policy.

Since the early 1980s the affluent have seemed, if anything, less inclined to help the poor than ever, and the gap between rich and poor in America has grown considerably. According to figures from the US census, in 1996 the top 20 per cent of families earned 49 per cent of US national income, up from 43.8 per cent in 1967. The bottom 20 per cent earned 3.7 per cent, down from 4 per cent in 1967.

But what has survived the era of Ronald Reagan in the USA, Galbraith insists, is a phenomenon he calls 'the liberal conscience'. The political theory of liberalism still enjoys a 'plausible social concern,' he says. By contrast, right-wing politicians dig for 'implausible justifications' for politics that reward corporations and the rich. 'The liberals still have a strong voice, a strong intellectual voice, and there's still a reflection of the liberal conscience in the US,' says Galbraith. Conservatives, on the other hand, using 'so-called supply side economics', have struggled to invent the notion that rewarding the affluent will result in social progress. 'The basic Reagan doctrine was that if you feed the horse enough oats, some will pass through the road to the sparrows,' he notes drily.

Galbraith himself is a piece of living history, and has every right to be looking back. He no longer keeps count of the number of books he has published (about 30) nor of the honorary degrees he has collected (about 50). The economist in him seems to take second place to the writer. 'I get nervously disturbed unless I do some writing every day,' he says. 'And my wife [Kitty, they have been married sixty years] would find me intolerable if I didn't.'

His books include a best-selling novel (*The Triumph*—a poke at the State Department and anti-communism) and range in subject from a memoir of his Canadian upbringing to a history of Indian painting.

Galbraith's most recent book is a collection of his letters to President John F. Kennedy. His next, to be titled *Name Dropping*, describes the political figures he encountered in a career that first took him to Washington in the early days of Roosevelt's administration.

Galbraith's higher education began at the Ontario Agricultural College (now the University of Guelph), from which he graduated with distinction in 1931. From Ontario, he moved to the University of California at Berkeley to study for a doctorate. He moved from there to Washington and briefly found work in the US Department of Agriculture, then went on to both Harvard and Cambridge universities. An assistant professorship at Princeton followed.

Having earlier worked for the Roosevelt administration, in the war years Galbraith headed the US office of wartime price control, after which he was a director of the US Strategic Bombing Survey. The USSBS had the task of assessing the impact of strategic bombing in first Germany and then Japan. Galbraith found the remains of German and Japanese cities a 'greatly sickening sight'. The SBS findings, of which he was a leading author, concluded that German war production, peaking in late 1944, had actually expanded through the bombings. And Japan would have surrendered, the SBS reported 'even if the atomic bombs had not been dropped.' These conclusions, that aerial bombardment did not win and barely even shortened the war, so outraged the US military that in 1948 its friends at Harvard delayed Galbraith's promotion to full professor in a bureaucratic battle that ran for a year.

Galbraith also worked for the State Department as administrator of economic affairs in occupied Europe, as an editor of *Fortune* magazine, and as adviser and speech writer to Democratic candidates and to Presidents Kennedy and Lyndon Johnson. Johnson would provide Galbraith with one of his favourite stories. 'Did y'ever think, Ken, that making a speech on ee-conomics,' the president asked him, 'is a lot like pissing down your leg? It seems hot to you, but it never does to anyone else.'

Kennedy appointed Galbraith ambassador to India. His collected *Letters to Kennedy* are about as far removed from the familiar tell-all biographies or nutty assassination conspiracies as it is possible to go. In one letter, Ambassador Galbraith teasingly informs the president that

Indian leader Jawaharlal Nehru was 'deeply in love' with Jackie Kennedy, and 'has a picture of himself with J.B.K displayed all by itself in the main entrance hall of his house.' That's as scandalous as they get. The letters begin in 1959, with Galbraith, riding high on the reception for *The Affluent Society*, emerging as an adviser on economics to the then Senator Kennedy. They end in November 1963, a week before Kennedy's assassination.

The letters confirm Galbraith's skill as a writer, his abiding reservations about the State Department as an institution and his dislike for Richard Nixon as a politician. They tell, in particular, of his prescient opposition to American military involvement in Vietnam, even before it had begun. 'It is those of us who have worked in the political vineyard and who have committed our hearts most strongly to the political fortunes of the New Frontier who worry most about its bright promise being sunk under the rice fields,' he wrote in a telegram on Vietnam policy in 1961. 'Keep up the threshold against the commitment of American combat forces,' he urged in 1962. 'I wonder if those who talk of a ten-year war really know what they are saying in terms of American attitudes.'

Galbraith has been celebrated by himself, and others, for an ego as large as his intellect—an 'unduly well-developed view of my own intellectual excellence,' as he calls it in his memoirs. 'Along with people who like to hear themselves talk,' he warns President-elect Kennedy, at the close of a longish epistle in November 1960, 'there are, unquestionably, some who are even more inordinately attracted by their own composition. I may be entitled to a gold star in both categories.'

For observers of American politics, there is one particular curiosity: Galbraith advising Kennedy to keep the radical right in perspective. The militia and white supremacist groups that inspired Oklahoma City bomber Timothy McVeigh are often presented as a modern American phenomenon; in fact, they were pre-dated by earlier militant tendencies. There are always about three million Americans, Galbraith observes, ready to follow the demagogue of the moment against law, decency, the Constitution, the Supreme Court, compassion, and the rule of reason. 'There's always in the American polity a margin just large enough to create concern for their violent heroics—but not enough ever to threaten the system.'

America's alliance with Britain, Galbraith told Kennedy in a 1962 memorandum, 'is one of the few substantial and workable alliances in the world'. He worried then that it would be lost if Britain got 'roped in' to Europe, and urged Kennedy, if he could not oppose Britain's joining the European Economic Community, to at least 'not shove unnecessarily'. In the 1960s and 1970s, he says, 'there was in my generation a concern about what was happening with British politics, almost equal to the concern with American politics'. British politicians, mostly on the left, were in weekly contact with their US counterparts.

Despite Britain's ever closer bonds with Europe, he says, the alliance has held. 'The British have a common language, a common literature, a close reciprocal understanding of politics and the political life, and, perhaps most important, a diminishing sense of national identity. There was a time when the British Empire was seen as unique, and the American Republic was seen as unique, and all this has diminished.'

by Tim Cornwell

Penalize the Bankers, Not the Workers

J. K. GALBRAITH

The economic and social development I would most like to see in the next century is one based firmly on what I have seen in the century just ending. It concerns poverty; there are two clear manifestations. In the great urban cities of the industrial countries there are still islands of deprivation, and especially so in the United States. Inequality is a basic condition.

The income gap must be narrowed and particularly by improvement of the condition of those now deprived. Nothing so denies enjoyment of life, and indeed liberty itself, as a total absence of money. Or some approach thereto. A rich country can guarantee an income to the deprived. If some do not work, so be it. The rich are also known on occasion to prefer leisure.

In the larger world there are the greatly impoverished populations. People are people: they suffer pain from starvation, lack of shelter, and from illness, wherever they are. As humans, they must be the subject of our compassion and of our help, our concerned action.

We must also now recognize that the end of colonialism left some countries with no government or with cruel, egregious, or incompetent governments that are in denial of any hope for well-being. In the years ahead there must be a procedure by which a strengthened United Nations suspends sovereignty in countries whose governments are destroying their people. We cannot in good conscience continue to accept such decades-long cruelty as that experienced, and still experienced, in the Congo. And elsewhere. And more generally, there must be willing and ample economic help from the fortunate countries to the poor.

There are other problems to be addressed. Capitalism still lends itself to instability deriving from reckless error as currently in Asia, in its nascent Russian form, in Latin America and, potentially, with an end to the Wall Street bubble in the United States. Our present remedial action bails out the bankers and industrialists who were most given to the causative insanity. It prescribes restraint on help to those most suffering from the disaster. So the oratory. So the International Monetary Fund, which rescues the bankers and the business executives responsible for the crisis, urges budget restraint at the expense of workers and the larger public. We must have the IMF, but in a more compassionate, more sociably equitable form. Here, needless to say, I yearn to see change.

Finally, all economists must with all concerned citizens take leave from everyday considerations to give thought and urge action that will end the greatest of dangers. That is from nuclear devastation. The most profound occurrence of this in the last century was the development of the means to destroy all life on the planet. That end is now fully available in the nuclear stockpiles, notably those of the United States and Russia. It awaits only the authority of some insane politician or his or her military delegates. We have already experienced the threat. No economist can take professional refuge from the omnipresent and overwhelming danger of nuclear destruction. Nor can anyone else.

Further Reading

Galbraith, J. K., *The Affluent Society* (4th edn., London, 1984).
—— *The Culture of Contentment* (London, 1992).
—— *The Good Society: the humane agenda* (London, 1996).
—— *Letters to Kennedy* (Cambridge, Mass., 1998).

Daniel Goleman

N THE BEGINNING THE WORD WAS IQ. EVERYONE BELIEVED IT WAS WHAT LED TO SUCCESS AT SCHOOL, AT COLLEGE, AND IN WORK. BUT THEN DANIEL GOLEMAN GAVE US EQ—EMOTIONAL INTELLIGENCE—AND HE WAS LAUDED BY EDUCATORS AND HAILED AS A MESSIAH BY THE BUSINESS WORLD.

When Goleman's book *Emotional Intelligence* first came out in 1995, it was an instant hit, catapulting its author headlong onto the lecture circuit. It has since sold over four million copies worldwide and been translated into 30 languages. Goleman has received hundreds of invitations to deliver the message of the book—that IQ is less important to how you do in life than what he calls 'emotional intelligence', a set of skills unrelated to academic ability. The book spoke to ordinary people, who felt released from feelings of inadequacy and society's fixation on IQ. Nor was this just another feel-good, self-help book full of well-meaning platitudes—Goleman had serious scientific data to back up his claims.

'I found that a new and compelling understanding of the emotions was being reached [by scientists]. For years, the study of the emotions had been ignored in neuroscience, cognitive science and psychology. Suddenly, we were starting to learn about the dynamics of the emotional centres of the brain, and I felt that there were implications for our personal lives and for important social problems, particularly the problems of young people,' recalls Goleman. He was then working as a science journalist for the *New York Times*, having left a fledging academic career years earlier to become an editor for the US journal *Psychology Today*.

Goleman saw the writing of *Emotional Intelligence* as more than a mere scholarly exercise. He hoped it might act as an antidote to the 'emotional malaise' that he felt had crept into society. This seemed to permeate everywhere, but was most pronounced among children—a dismal fact testified to by several school shootings in the USA in the 1990s. Goleman believed that by learning to recognize emotional intelligence, we could improve almost all aspects of our lives—from school to work to our personal relationships, and so become happier, healthier people. This rosy message stemmed not from naive hopefulness, but from a new understanding of human emotions emerging from neuroscience.

Emotional intelligence is nothing mysterious or magical, it is something we all recognize. It is typified by people who seem at ease with themselves and with others—people who get ahead but also get along with all kinds of people. Goleman identifies five main 'domains' of emotional intelligence: the first is 'self-awareness', the ability to recognize your own emotions, to know your strengths and weaknesses and to generate a sense of self-worth. The second is 'self-regulation', the ability to control your emotions rather than allowing them to control you. The third is 'motivation', the strength of will needed to achieve your goals and to pick yourself up after a fall. While these first three areas concern your own emotions, the remaining two, 'empathy' and 'social skills', relate to other people's emotions, the ability to recognize them, and to nurture relationships or inspire others.

His books are grounded in examples from all walks of life of both emotional intelligence and emotional stupidity. In one chapter he describes the New York bus driver who soothes an irritable crowd of passengers on a stuffy bus with his upbeat and jovial banter. He later contrasts this with a cab driver in the same city who feels compelled to vent his frustration at the traffic by shouting obscenities out the cab window. And as an illustration of how IQ and emotional intelligence have little to do with one another he tells the story of the straight-A student who stabs his physics teacher for giving him a B.

Goleman brings insights in neuroscience to bear on this common-sense understanding of the emotions, drawing together hundreds of pieces of scientific research to explain how two parts of the brain—the 'emotional brain' and the 'rational brain'—link up to form the neural circuitry responsible for emotional intelligence. An intricate dance between the amygdala, the ancient emotional centre responsible for our primal impulses, and the prefrontal lobes of the newer neocortex, dictates how we behave in everyday situations.

'IQ is based on the neocortex alone, and the neocortex can do that [largely] without the emotional centres,' explains Goleman. 'But when we talk about emotional intelligence, we're talking about integrating a wider sweep of the brain, both limbic and cortical areas.'

The amygdala is responsible for the rapid 'fight or flight' response in people faced with a difficult situation. It creates an 'emotional short-circuit', taking over the entire brain in an instant. But in most situations the prefrontal lobes are able to temper the amygdala's primal response with reason. If the prefrontal areas do a good job, a person might quickly defuse a tense situation with humour. If not, the result could be an uncontrollable rage.

Emotional Intelligence inspired a host of educational programmes in social and emotional development around the world. Educators rallied to Goleman's cry that something could be done to reverse the current trend among children to be emotionally troubled, unruly, depressed, and prone to worry. But while he had expected teachers to take an interest, Goleman was taken aback by the book's wider impact.

'I was very surprised to see that I'd got as much, if not more, interest from business people as from educators,' he says. As invitations to speak began to flood in from companies all over the world, Goleman had to give up his day job at the *New York Times*.

'I had never had much business experience or interest in business, and I actually grew up with a slight disdain for business people, which I think many academics feel,' he admits. 'I had a stereotype that was quite erroneous—I find the business world a more congenial universe to operate in than I had anticipated.'

His latest book, *Working with Emotional Intelligence*, explains how emotional intelligence is at least twice as important as IQ or technical skills in determining how well a person will perform on the job. In it he warns businesses that they are wasting billions of dollars every year on training programmes that are destined to fail.

Over the years, the business world has come to believe that IQ and personality tests can help them pick the right people to hire. But it was one of Goleman's own professors at Harvard, David McClelland who pointed out over twenty-five years ago that while IQ is a good predictor of what job a person can get, it is a very poor predictor of how well they will do once employed. This, he suggested, was because IQ measured the wrong set of skills.

In the intervening time, companies have followed McClelland's advice and have looked at what distinguishes 'star performers' from average workers. Time and again, says Goleman, it is emotional intelligence competences, such as initiative, self-confidence, relationship abilities, and the drive to improve, that these people excel in.

While on average these skills are twice as important as cognitive skills, they become ever more important the further up you go in an organization. At the top, they can account for as much as 85 to 90 per cent of the characteristics distinguishing stars from plodders.

A well-known test of emotional intelligence and its impact on our lives is the so-called 'marshmallow test'. A marshmallow is placed in front of a child aged four, who is told that it is his, but that if he can wait while the adult goes out of the room for a while, then he can have two marshmallows on her return. One such study was conducted at Stanford University in the 1960s. When the children were later tracked down as adolescents, those who had been able to delay gratification at age 4 were found to be more socially competent, less likely to cave in under stress and more self-assertive than those who gave into temptation and grabbed the marshmallow. They also achieved considerably higher grades upon leaving school than their more impulsive peers.

While this example paints a picture of emotional intelligence as something more or less fixed throughout life, Goleman firmly rejects that conclusion. 'There is probably a Bell curve for emotional intelligence just as there is for IQ, but the difference is that the emotional brain is very plastic—it continues to reshape itself through repeated experience. That is one reason I'm against evaluating people for emotional intelligence the way we do for IQ, particularly in childhood,' says Goleman. 'People might treat an EQ score the same way as an IQ score, and say, "Well, she's not so good at this," and it becomes a self-fulfilling prophecy.' Emotional intelligence rises throughout each decade of life—in other words, says Goleman, people mature.

One important insight from neuroscience is that the emotional brain learns in an entirely different way from the rational brain. While a classroom setting and textbooks may be appropriate for learning technical skills, they are next to useless for learning how to behave in a more emotionally intelligent fashion. The appropriate model of learning is one of habit change, learning new skills through practice and repetition. 'It's not enough to just read about emotional intelligence or hear about it—that is actually quite ineffective when it comes to our emotional repertoire. In fact, companies have been wasting billions of dollars on training and development programmes that don't pay, because they follow the wrong model of learning,' asserts Goleman.

During Goleman's own upbringing, academic success was placed in especially high regard. Both his parents were academics working in small colleges in his native California, and he had always assumed that he too would become an academic. After a Ph.D. in psychology at Harvard, he eventually fulfilled that destiny, taking up a position on the faculty as a visiting lecturer. And though he has not formally worked in academia for many years, Goleman says that he now considers himself an 'applied academic'.

His early research focused on the physiological basis of meditation, and after some trips to Asia he wrote his first book *The Meditative Mind*, which examined how meditation could act as an antidote to stress. Later, while working at *Psychology Today*, he wrote *Vital Lies, Simple Truths*, this time analysing the physiological basis of self-deception. 'I've always been intrigued by the role of attention in mental life. Meditation is a sustained attempt to retrain attention, while you could say that self-deception is a pathology of attention,' says Goleman, attempting to explain a common thread that has run through his work.

Another common thread is his interest in understanding behaviour by looking at the neural processes that underlie them. 'My whole career as a psychologist has been on that cusp,' he says. 'My first research was psychophysiological—that's one of the growing tips of psychology today. Increasingly, people are eager to make cross-connections between behaviour and the brain mechanics that underlie them.'

Goleman's flair for tapping into the popular imagination was evident early in his career. While at Harvard, he taught a course on the psychology of human consciousness. 'It was immensely popular—it was so big that they had to reschedule it to the largest auditorium!' recalls Goleman. This empathy for other people's interests may be an aspect of his own emotional intelligence. And as he describes his own success and popularity, he does so without undue modesty, but also without arrogance. If he has any weakness, it is perhaps a slight detachment, a reluctance to give much away about himself. Among the many anecdotes in his books on emotional intelligence, very few are drawn from his own life and experience. Perhaps he is just a little shy.

by Ayala Ochert

Winning the Battle for the Human Heart

DANIEL GOLEMAN

'You can plan for a hundred years, but you don't know what will happen the next moment,' an Indian sage once told me. With that caveat in mind, let me speculate about what I see as the coming battleground for the human heart, and thus the future of humanity. I predict that developed societies will expand the mandate of schools to include an education of the emotions.

The human dilemma, as I see it, revolves around a paradox of neural architecture: the brain has evolved in a fashion that grants the emotional centres immense power in dictating our reactions. And so people with even the greatest sophistication and most modern technology can respond with a primitive emotional repertoire. Einstein captured the dilemma aptly when he said, 'The splitting of the atom has changed everything save the way men think, and thus we drift toward unparalleled catastrophe.'

The signs of that drift proliferate. True enough, for the time being the great powers have averted an outright nuclear confrontation. But small-scale wars abound, as do ethnic battles and simmering inter-group hatreds. And within the neighbourhood and the family, violence has a shockingly high prevalence.

As I write this, America, that bell-wether culture, reverberates with the shocking news of high school students massacring their fellows with automatic weapons, all-too-easily obtained. Those events, sad to say, fulfil another prediction I made in the early 1990s, as I was writing my book *Emotional Intelligence*.

That earlier prediction was based on massive data, most telling perhaps was a study of a representative sample of close to 3,000

American children and teenagers evaluated by their parents and teachers. The data showed that over a decade-and-a-half, America's children declined on average in basic abilities to handle themselves, their feelings, and their relationships: they become, for instance, more impulsive and disobedient, more irritable and angry, more anxious and lonely. The decline was on 42 measures of self-management and social skill. They went up on none.

Along with that invisible decline came more troubling headlines: sharp rises in rates of homicides, suicides, and rapes among teenagers. These and other more visible indicators of a lack of emotional intelligence among America's children have begun to surface in developed nations around the world.

Children appear the unintended victims of two enormous forces at large in the world, one economic, the other technological. The ratcheting up of competitiveness in the global marketplace means that today's generation of parents has to work harder and longer to maintain the same standard of living as their parents—so they have less free time to spend with their own children. Mobility means that fewer families have relatives nearby to help with day care, and too many families live in neighbourhoods where they are afraid to let their children play outside unsupervised.

So today's children spend much time on their own. One way they fill that time amounts to an unprecedented experiment with the world's youth: never before in human history have so many children spent so much time staring at a video monitor. Whatever is on that monitor, it means they are not out playing with other children. And the way we have passed on emotional and social skills is in life: from our parents, relatives, neighbours, peers.

This transfer of basic life skills simply does not happen as well as it used to. Schools, however, offer a vehicle for society to be sure that each new generation learns the basic arts of life—impulse control, managing anger and anxiety, motivation, empathy, collaboration and working out disagreements positively. There are already schools where the curriculum has been expanded beyond the basics to include lessons in these essential skills. The best teach a full spectrum of abilities, from the earliest school years through adolescence, with developmentally appropriate lessons in each year.

The results are highly encouraging: children not only get better at self-management and handling relationships, but the numbers of fights and violent incidents decline—and academic achievement test scores rise. If we are to change that drift towards catastrophe, every child should have access to these lessons.

In short, my prediction holds that one day every schoolboy and girl will be taught these pragmatic arts of living well, along with the traditional academic basics. Empathy will hold as valued a place in the curriculum as algebra.

Further Reading

Goleman, Daniel, *Emotional Intelligence: Why It Can Matter More Than IQ* (London, 1996).

—— *Vital Lies, Simple Truths: the psychology of self-deception* (London, 1997).

Stephen Jay Gould

A SMALL ANECDOTE SAYS A LOT ABOUT STEPHEN JAY GOULD. I ASK HIM ABOUT THE PHILOSOPHICAL IMPLICATIONS OF HIS RECENT BOOK *LIFE'S GRANDEUR*. HE TELLS A STORY ABOUT REVIEWING ONE OF PHILOSOPHER/PHYSICIST FRITJOF KAPRA'S BOOKS ON COSMOLOGY: 'HE'S A CALIFORNIA MYSTICAL TYPE, I MEAN A NICE GUY. WHAT I SAID IN MY REVIEW OF HIS BOOK WAS THAT I'M A HOLIST LIKE HIM IN SOME SORT OF PHILOSOPHICAL SENSE, BUT I'M A NEW YORK HOLIST, NOT A CALIFORNIA HOLIST. I BELIEVE IN INTELLECTUALITY, NONE OF THIS CALIFORNIA TOUCHY FEELY STUFF.'

Although for the last thirty years Gould's academic home has been the WASPish elegance of Harvard University, Gould remains a cosmopolitan New Yorker at heart. He was born in 1941 ('the year Ted Williams hit 400 for the last time'—baseball is a Gould obsession), in the borough of Queens, the son of second-generation Jewish immigrants.

He picked up his appreciation of life's variety and his love of dinosaurs early, when he visited the New York Museum of Natural History with his father: 'Journalists love that story, but to me it's banal, because it's so common among palaeontologists. More than any other profession I've ever come across, we were childhood enthusiasts.'

He was educated in the New York public school system, and has fond memories of his Jewish and Irish teachers. In school he was also taunted as a 'fossil-face', a less pleasant memory. His first degree was at Antioch College; his Ph.D. at Columbia in New York. At Harvard since 1967, he is now professor of geology and a curator of the Agassiz Museum of Comparative Zoology. Despite his love of dinosaurs, his palaeontological speciality has been land snails, and he still goes out and chips away at bits of rock, although where he finds the time to do so is difficult to imagine.

For Gould is one of the most prolific essayists of our times. For the past eighteen years he has written monthly essays for *Natural History* magazine, many collected in his books which invariably sell hundreds of thousands of copies. His recipe for such productivity is simple: 'Pull out the phone plug and don't write drafts. I maintain the old-fashioned skill of writing outlines, so I know what I'm going to write when I start.'

His essays are extraordinarily eclectic. In just one recent essay, for example, he ranges across subjects that include an extinct species of South African antelope, the Nazi eugenics programme and nineteenth-century women naturalists. He delights and sometimes irritates his readers with his fascination for trivia, and the way he uses obscure examples to make his points. But he is an outstanding popularizer of science, who seems able to convey scientific complexities to non-scientists without sacrificing accuracy to sloppy metaphors.

His one stylistic fault is a tendency to prolixity. Critics sometimes quote passages where it is difficult to escape the conclusion that he is indulging his pleasure at his own intelligence. One reviewer even accused him of 'vulgar cleverness'.

In the academic sphere, despite universal respect for his scholarship, some critics have been harsher. Evolutionary biology has always been a contentious subject, for the simple reason that so much is unknown and unknowable. That leaves plenty of space for disagreement over theory, and, as befits a Jewish intellectual, Gould is nothing if not contentious. The results of his efforts are usually productive, but his critics claim that he takes extreme positions, constructs straw men and is unnecessarily combative. He is careful to stress his areas of agreement, for example, with arch-opponent and rival popularizer Richard Dawkins, and it is clear that he is sensitive to these charges.

The central debate of his career has been about the long-term dynamics of evolution. In 1972, together with the zoologist Niles

Eldredge, he launched the theory of 'punctuated equilibria', according to which species stay much the same over long periods of time, but then undergo rapid bursts of development during which new species are formed. It is in these rapid bursts that the most evolutionary change takes place, rather than gradually over aeons, as the prevailing view had it. Gould's evidence for these radical claims comes primarily from the fossil record, which, he says, contains no evidence for gradual changes within a species.

Punctuated equilibrium theory has various revolutionary implications for evolutionary biology and the theory has provoked a vigorous response from defenders of the previously prevailing consensus. But perhaps most shocking is Gould's downplaying of the importance of natural selection—the theory that animals evolve largely as a result of adapting, over generations, to their environment, which 'selects' those characteristics that best enable the animals to survive and reproduce.

Gould insists, rather, that contingency plays an important role within evolution. According to the traditional view, natural selection constantly creates gradual evolutionary change by favouring the tiniest selective advantages. But if the fossil record indicates long periods of evolutionary stasis, natural selection cannot be this kind of creative force.

A further implication of the older paradigm is that if we knew what the selective forces were, we might be able to predict which evolutionary changes would occur. But Gould points out that evolution can only work with what it has to hand, so much of evolutionary change is accidental: if a successful individual happens to have a particular number of toes, for example, then all its descendants will have that number of toes, even though having that number of toes confers no particular advantage.

In pressing this point Gould sometimes goes too far, laying himself open to the charge that he does not believe in natural selection. But he insists that his work is an extension of Darwinism, not a challenge to it. 'The overarching theme of my work is that you can't have an adequate evolutionary theory merely by looking at what's happening in local populations at the moment and extrapolating its adaptationist and gradualist style of change to the entire history of life. That's what Darwin wanted to do . . . My examination of the fossil record requires that you develop a body of independent theory for large-scale events that take long periods of time . . . Natural selection is a powerful, beautiful theory, and it's correct. I just don't think it's fully adequate, or close to fully adequate.'

Gould is also well known for his criticisms of sociobiology, an academic subject which holds that the way humans behave and organize themselves socially can be explained as biological adaptations to evolutionary forces. I ask him what he thinks about recent efforts to rehabilitate sociobiology, under the new title of evolutionary psychology, with its additional argument that modern human behaviour can be explained as a legacy of the kind of traits humans needed to survive two million years ago?

The response is prompt and gently scathing. 'No one ought to have any objection to the proposition that evolution should be able to offer profound insights into the nature of behaviour. The problem with the old sociobiology and the new evolutionary psychology is that they're very naive pan-adaptationist theories, which rely entirely on the adaptationist component of Darwinism.'

'The failure of the old sociobiology was that it tended to look at just about anything as adaptive now, or at least maintained for adaptive reasons. Today's evolutionary psychologist will say: "No, we recognize that's wrong, there are many evolutionary coded behaviours, like aggression, which are profoundly non-adaptive now, but, when they arose on the African savannas, they were adaptive."'

'Now in a sense that is a more sophisticated insight, I'll grant them that. On the other hand, as a scientific proposition it's even worse than the old-time sociobiology, because at least the old sociobiological theories could be tested. If you say it's adaptive now, OK, go out and see if it does increase reproductive success. But the minute you say, "it was adaptive on the African savannah", it's not a scientific theory any more. There's no way to test that proposition . . . language doesn't fossilize, kinship doesn't fossilize. You're reduced to speculative story-telling about hunter-gatherers on the savannah.'

A related aspect of Gould's revised Darwinism is that there is no progression within evolution. We do not necessarily evolve in the direction of greater and greater sophistication. Rather, Gould argues that the main mode of life on this planet has been and will remain bacterial. But if life started with bacteria, and progressed through sponges and dinosaurs to *Homo sapiens*, surely that's progress? Only from a culturally determined viewpoint, and based on a misapprehension of the nature of reality, says Gould.

He fulminates at length against the evolution-as-progress interpretation of Darwinism, to which Darwin himself succumbed only reluctantly, and which was imposed due to the nineteenth-century ideology of progress, and man's self-importance. Evolution has always tended to be depicted as a ladder ascending to humans, or a series of 'ages', always increasing in complexity. Yet the most common value of the distribution of complexity remains firmly anchored at the simple bacterial end. Increased complexity in some species arises randomly. Because it is impossible to get less complex than bacteria, random motion will, over time, produce an expansion at the other end of the complexity distribution. But we should always remember that we are produced by random motion; we are not an inevitable result of natural selection operating in favour of increased complexity. For Gould this is the completion of Darwin's revolution, the dethronement of humanity from its self-centred view of its own importance.

Gould is, of course, part of the great tradition of liberal humanism, and proud of it. But his emphasis on diversity and rejection of the ideology of progress strikes resonances with postmodernism. It is a label which, true to type, Gould is reluctant to wholeheartedly embrace: 'Of course, there are aspects of my work which are consistent with postmodernism. But no liberal scientist has any time for the notion of the relativity of truth. Of course, science is socially embedded, all science is done in a social context. But there is an external reality out there, and we do get a better approach towards it as we proceed through the history of science.'

by David King

Unpredictable Patterns

STEPHEN JAY GOULD

The turnings of our centuries may bear no relationship to any natural cycle in the cosmos. But we construe such artificial transitions as occasions for taking stock, especially at the centurial boundaries that have even generated their own eponymous concept of cyclical angst—the *fin-de-siècle* phenomenon.

At the first millenial experience in Western history, Europe feared all the gory prophecies of Armageddon as recorded in *Revelation, chapter 20*; this second time round, we focus our worries on what might happen when computers misread the great turning as a recursion to the year 1900. (I am amused by this diminution in the quality of anxiety for the two documented transitions.)

Nonetheless human futures are unpredictable and it is futile to think that past trends will forecast coming patterns. The trajectory of technology might offer some opportunity for predicting the future—as science moves through networks of implication, and each discovery suggests a suite of subsequent steps. But even the 'pure' history of science features unanticipated findings, and must also contend with nature's stubborn tendency to frustrate our expectations—factors that will cloud anyone's crystal ball.

Moreover any forecast about the future must consider the incendiary instability generated by interaction between technological change and the weird ways of human conduct, both individual and social. How can the accidents that shaped our past give any meaningful insight into the next millennium? Pasts can't imply futures because a pattern inherent in the structure of nature's materials and laws too often disrupts an otherwise predictable unfolding of historical sequences.

Any complex system must be constructed slowly and sequentially, adding steps one (or a few) at a time and constantly coordinating along the way. But the same complex systems, once established, can be destroyed in a tiny fraction of the necessary building time—often in truly catastrophic moments. A day of fire destroyed a millennium of knowledge in the library of Alexandria and centuries of building in the city of London. The last blaauwbock of southern Africa and the last moa of New Zealand perished in a momentary shot or blow from human hands but took millions of years to evolve.

The discordance between technological and moral advance acts as a destabilizing factor. We never know when and how these episodes of destruction will fall, sometimes purging the old to create a better world by revolution, but more often simply cutting a swathe of destruction and requiring a true rebirth from the ashes of old systems as life has frequently done—in a wondrously unpredictable way—following episodes of mass extinction.

Thus, I am not sanguine about possibilities for predicting the future. Even to anticipate a single scientific breakthrough in the next century and then to speculate about its implications for human life carries our hopes for prophecy too far.

Further Reading

Gould, Stephen Jay, *Eight Little Piggies: Reflections In Natural History* (London, 1993).

—— *Life's Grandeur: The Spread Of Excellence From Plato to Darwin* (London, 1996).

—— *The Mismeasure of Man* (London, 1997).

—— *Questioning the Millennium: A Rationalist's Guide To A Precisely Arbitrary Countdown* (London, 1997).

Susan Greenfield

WHAT DO YOU THINK I WOULD KNOW OR DO IF THE PROBLEM OF HUMAN CONSCIOUSNESS HAD BEEN SOLVED—IF WE UNDERSTOOD EXACTLY HOW WE EXPERIENCE THE COLOUR GREEN OR THE SMELL OF COFFEE BREWING OR THE PAIN OF A GRAZED KNEE?'

The query hangs in the rarefied air of the Royal Institution, where the director is the bright and breezy Susan Greenfield, leading researcher and media darling.

'Would I be able to climb into your brain?' she goes on. 'Would I be able to share your consciousness, in which case you would be denied your innate individuality? Would I be able to build consciousness? What exactly could I do with the knowledge of how human consciousness happens?'

The riddle of consciousness has troubled Greenfield since she was a teenager and her mother told her that neither of them would ever know what the other felt when they saw the colour red. But why is the puzzle so difficult to solve? Painstakingly, she explains: 'The big problem is asking the right questions,' she says. 'The ultimate question is how do physical events within the brain, across synapses and neurons, translate into subjective experience? How does, for instance, changing the availability of the chemical serotonin in the brain translate into a subjective feeling of well being?' How, in short, does the grey mass of brain tissue generate our sense of self?

Greenfield's quest for an answer has taken her from philosophy to psychology to neurobiology. 'I came to Oxford as an undergraduate to study for a philosophy, politics, and psychology degree,' she explains. 'Reading the classical authors fuelled my interest in the mind— philosophers were the first to ask questions about how we subjectively experience sensation. But studying psychology got me interested in how the brain physically works.' She switched to a degree in experimental psychology to get closer to the problem. 'It was a whole new world.' Studies in disciplines as varied as neurochemistry, animal behaviour, and computation followed—anything that could shed some light on how the brain works.

Greenfield is as close as any of the many scientists investigating consciousness to finding at least part of the answer. At the moment, as professor of synaptic pharmacology at the University of Oxford (she splits her time between Oxford and her London post), she is exploring connections between two diseases of the brain; Alzheimer's and Parkinson's. Conventional wisdom dictates that there is no direct link, but Greenfield has yet to be convinced. 'My own view is that the two diseases, although they exhibit different symptoms and are treated by different medication, actually share common factors,' she says. 'What we think happens, in both cases, is an aberrant form of development. Because, for some reason, the brain cells of sufferers think they should be growing again, they bring into play mechanisms that are fine when a foetus is developing but that are pernicious in old age.'

Last year Greenfield, with three colleagues—David Vaux, Nick Rawlins, and Martin Westwell—launched a company, Synaptica, a spin-off from the university. Synaptica's goal is to identify a drug which could be used to treat both Alzheimer's and Parkinson's.

'We are trying to discover the critical chemical that is common to, and might be responsible for, the cells that are lost in both Alzheimer's and Parkinson's,' says Greenfield. 'If we could discover it and characterize its action, then we could design a drug that would tackle both diseases, a medication that would arrest degeneration of the brain.'

'If you could get it right, the market would be huge,' she adds. 'It is predicted that by 2050 in the United States alone, up to 14 million people will suffer from Alzheimer's. And it is not just the patients who are affected. Arguably, Alzheimer's disease is more stressful for relatives, because the patient is unaware of what is happening, whereas for the carer it is devastating to watch the decline into senility of someone you love.'

A better understanding of the brain would also help Greenfield unravel the enigma of consciousness. But there are many barriers to hurdle first. She is enthusiastic about the relatively new diagnostic tool of brain imaging—whereby, using a scanner, scientists can produce a picture showing which parts of the brain are used for different tasks. 'With brain imaging, people at last have a window into the functions of the brain which was never accessible before.' But she worries that brain imaging is still a relatively crude tool. On its own, she says, it 'cannot inspire a theory of how the brain works. All it tells you is that if you do this or that, different bits of the brain light up on a scan—it doesn't tell you why or how or what that means.'

Another shortcoming of brain imaging is that it only monitors events in the brain which last longer than one second. 'We know that brain cells work on the scale of thousandths of seconds,' says Greenfield. 'In terms of finding what in the brain matches up with different moments of consciousness, we are going to need brain imaging to be much more sophisticated.'

In fact, perhaps surprisingly for a woman so enthusiastic about life in general, Greenfield is downbeat about many of the grand claims being made by scientists at the end of the twentieth century.

Of the geneticists' hope that we shall soon be able to pinpoint and modify genes for human behaviour; genes for shyness, for instance, or for intelligence, she is roundly dismissive. 'I think people may be disappointed by the mapping of the human genome,' she says. 'Even if you know what a gene is for, you don't know how that relates to all the functions of the brain. To an order of magnitude, you have 1 million genes and 100,000,000 million brain connections. So I think there might be an over-expectation of what molecular biology can deliver. It is not conceptually viable to find the gene to make people fall in love with each other.'

Nor will the artificial community succeed in their goal of building a conscious robot, she says. 'The idea is ridiculous. Consciousness entails an interaction between the brain and the body, trafficking a myriad of chemicals between the two. To reproduce that, you would have to build a body with a whole range of chemicals, and the three-dimensionality of the brain would have to be preserved to the very last connection.'

But if scientists don't have the answers, nor, according to Greenfield, do the philosophers. She has worked with philosophers to debate and construct different theories of consciousness, and has co-edited a book of debates between philosophers and scientists called *Mindwaves*. 'To be honest, I can't see how philosophy on its own is going to explain how the physical brain is translated into subjective experience,' she says. 'I think philosophers' role—and I shall probably offend them hugely by saying this—is to provide checks and balances on any theories a scientist might come up with.'

Her own, admittedly incomplete theory, is that consciousness emerges from the activity of groups of nerve cells in the brain. In her book, *Journey to the Centres of the Mind*, she writes: 'the more I heard, read and thought about it, the more it seemed incredible that mere molecules could in some way constitute an inner vision, idea or emotion, or—even more astounding—that they could generate the subjectivity of an emotion. Yet consciousness is continuous with the brain's activity and must emerge from it.'

'I don't think the problem of consciousness will be solved,' she says. 'Just because we realize that vitalism is outdated, just because we can couch things in physical terms, it does not mean that we will be able to solve the hard problem of consciousness.'

Greenfield's career has been primarily based at Oxford, with spells in France and the United States. I ask her what it is like to be a female scientist, and she retorts that since she has never been a male scientist, the question is difficult to answer. 'When I was younger, I did not see [being a female scientist] as a problem,' she says. 'I got rather impatient, a bit like Margaret Thatcher used to get with people about being a woman in politics. Paradoxically it was only as I got more senior that I realized there was a problem. Previously, when I was patronized and ignored, I thought that it was because I was junior.'

'Some women, especially the younger ones, ask what to do [about discrimination]. If they protest, they get labelled "neurotic" and "hysterical". My advice is that women should network. If you feel, as a woman scientist, upset or troubled by something, go and compare notes with another woman. She will tell you if you are just being silly, or if she has experienced the same thing. It is a way of letting off steam but at the same time channelling it positively rather than being labelled as a prickly character who has not got a sense of humour.'

Greenfield also wants to see women scientists earn higher salaries. 'It is something that has to change,' she says. 'What saddens me is the women themselves: they feel like impostors and that it is just a fluke that they have reached the positions they are in. Women are easily put down by men and will back down in an argument. We need a change of mindset to help women attain a sense of their own worth, and to help them deal with confrontation.'

Greenfield is angry, too, about how women scientists still have to choose between children and a high-flying career. 'Many women, in the prime time of their research, will take time off to have children,' she says. 'And it is very hard to come back. It should be the norm that having a child is not going to jeopardize your scientific career. At the moment women are being given a horrible and unfair choice.'

Greenfield, aged 48, who married Oxford chemistry professor and millionaire Peter Atkins in 1991, is clear about her own choice to remain childless. 'In retrospect, I was never the maternal type,' she says. 'I had a small brother and I was disabused early on about what babies are like.'

In fact, being female has made Greenfield stand out: she is frequently asked to appear in the media as a high-calibre scientist who is not just articulate and lively but also a woman in a male-dominated profession.

The first woman to give the renowned Christmas lectures at the Royal Institution and author of several popular books on how the brain works, her next big media project is a six-part BBC television series scheduled to be aired in April 2000. Provisionally called 'The Secret Self', it will develop her ideas about how consciousness and the emotions are formed.

For one so pessimistic about the breakthroughs expected of science in the near future she is deeply enthusiastic about helping the ordinary person understand scientific issues. 'In these programmes I can unleash on an unsuspecting nation my ideas on how the mind works,' she enthuses.

by Alison Goddard

George Orwell's 1984 ... in 2084

SUSAN GREENFIELD

A whole range of sure predictions can be made in the field of biomedicine. Each of these are often made independently of the others, but, when pieced together, I think they may well result in a surprising scenario.

Certainly, everyone could safely predict that, very soon in the next century, the human genome will be mapped; the problem, of course, lies in what will be done with the information. My fear is that the concept of the gene 'for' a particular human trait will be over-interpreted, not least when it comes to the function of the human brain. Already we have heard claims of genes for criminality and for homosexuality and yet we know that even if every gene in the body contributed to a brain connection the match between genes and brain connections would be far from exact; in fact one would still be out by a factor of 10^9.

Of course, some functions of the brain are capable of being modi-fied by genetic manipulation, and will be in the future. But such modifications will be crude at the beginning, lacking precision. Because of this imprecision it will just not be possible, fortunately, to engineer all homosexuals or all criminals out of existence. Nonetheless, the genetic modification we do undertake in the future will inevitably lead to a reduced gene pool, depleted, for instance, of genes such as those linked to diseases like cystic fibrosis. Knowledge of the genome will, of course, also mean genetic screening—monitoring for traits in people that might prejudice insurance policies.

The monitoring and manipulation of the phenotype (how a gene is expressed and observable) will be increased with the use of nanotechnology—small devices that detect and interfere with reactions throughout the body.

On a grosser scale, brain imaging, currently already the darling of neuroscience, will become increasingly commonplace and refined. Not only will it be far cheaper and easier to image someone's brain, but the timescale will mean that we will be able to correlate transient conscious states with physical and chemical events in the brain in a much more accurate way than can now be envisaged.

And as leisure time increases, then the current tendency, especially of the young, to interact with the artificial worlds offered by virtual reality helmets will increase. Just as nowadays we tend to 'escape' into a novel, so increased leisure time and frustration at being less than perfect might inspire some to escape into an artificial way of life. The risk is that the attention span and imagination needed to read a novel and so carefully cultivated through our childhood until now, will no longer exist: we may end up a society of restless, unimaginative individuals.

Moreover, since we know that the human brain is very sensitive to post-natal experiences, we may discover that standardizing the stimulation delivered to a child's brain—the artificial life offered in a VR world—may result in the actual standardization of an entire generation.

Drawing together all these individual predictions I anticipate that the individual human being will be monitored and manipulated in the next century on a scale never before contemplated—from their genes and molecules through to their mental worlds.

It may therefore turn out that 2084 is a much more accurate prediction of the 1984 scenario than that original date. We may end up with a society of increasingly standardized individuals, drawn from a far smaller gene pool. One risk is that we may obliterate forever the possibility of original thinking, the kind of thinking that results in paradigm shifts.

We are at a critical time. We possess the science and technology to either make everyone more standard or, to create for the first time, a world where we can really understand, appreciate, respect, and celebrate the diversity of our individuality. The choice is ours, but time is running out.

Further Reading

Greenfield, Susan, *The Human Brain—A Guided Tour* (London, 1998).
Rose, Steven, *Lifelines* (London, 1997).

Lynn Margulis

'D RATHER SHARE A PLATFORM WITH ATTILA THE HUN,' SNAPPED RICHARD
DAWKINS, OXFORD ZOOLOGIST AND BEST-SELLING SCIENCE WRITER, WHEN HE
WAS ASKED BY CAMBRIDGE STUDENTS TO SPEAK ALONGSIDE THE AMERICAN
BIOLOGIST LYNN MARGULIS. SHARP, EVEN BY DAWKINS'S STANDARDS, AND
THIS ANTIPATHY TOWARDS MARGULIS, WHO IS PROFESSOR IN THE DEPARTMENT OF
GEOSCIENCES AT THE UNIVERSITY OF MASSACHUSETTS, IS SHARED BY MANY
SCIENTISTS.

Yet there are many other academics who speak very highly of the
woman the journal *Science* dubbed 'Science's unruly Earth mother';
American sociobiologist Edward O. Wilson, for instance, calls her 'one
of the most successful synthetic thinkers in modern biology'. What is it
that makes this 62-year-old mother of four so controversial—a sort of
Noam Chomsky of the life sciences?

I met her in a cluttered curator's office at New York's American
Museum of Natural History. Her forthrightness was immediately
apparent—there is nothing politically correct about Margulis, who
particularly resists attempts to pigeonhole her as a role model for
young female scientists. 'People constantly ask me to talk about "the
feminist view of evolution",' she sighs.

'I hate the words role model. I spend most of my time trying to get
back to doing science. It's extremely hard, and when I am labelled a
woman's role model it is harder.'

Even outside science, Margulis has no use for women's studies as a
separate discipline. 'I think it's a bunch of phoniness. It's an excuse for
people who can't make it in real study. There's no women's history,
there's only history. My daughter calls me a "closet feminist", but I am
not a feminist at all.'

Strong views—but it is not these that get under the skin of neo-Darwinists like Dawkins. What irritates them is Margulis's radical expansion of Darwin's theory of evolution and her effervescent championing of the controversial Gaia hypothesis of the British scientist James Lovelock. The Gaia theory (named after the ancient Greek goddess of the earth) suggests that earth can be considered as a single, self-regulating, 'living' system. It irritates the majority of scientists who regard it as both unnecessary and 'mystical.'

In their view, it is best to regard living creatures (plants, animals, and microbes) and the non-living environment (rocks, oceans, volcanoes, etc.) as independent, though interacting systems that will ultimately be explained at the level of atoms, molecules, and genes by the laws of physics and chemistry alone. 'Natural selection,' Darwin's famous idea, thus consists of living creatures competing with each other, under the conditions of a dead environment, in order to leave maximum numbers of offspring. Those that are fittest survive and reproduce, whereas the rest die off.

Margulis explains: 'To me Gaia has no goddess implications at all. Gaia is the system at the surface of the earth that gives the planet properties of life.' These 'properties of life' include the earth's ability to regulate atmospheric gases like oxygen and to modulate surface temperature and the acidity and alkalinity of the oceans so that life flourishes. 'The Gaian view is useful because it leads one to ask questions which one would not otherwise ask, such as why has oxygen in the earth's atmosphere been relatively stable at 20 per cent for millions of years? Why is the atmosphere of lifeless Mars and Venus very unlike that of the Earth, with very little oxygen?'

Why on earth would 'adaptation' to a low-oxygen environment have produced oxygen-loving animals and plants? 'It makes no sense on the old-fashioned adaptationist paradigm, but it is entirely understandable from a Gaian point of view,' says Margulis.

None of this would have been enough to provoke scientists to attack Margulis, had it not been for the fact that, in the past decade,

Gaian science has scored notable successes and attracted talented scientists (whether or not they use the goddess's name or prefer the more fundable 'earth system science'). Even more important is that growing evidence for Margulis's key scientific idea, Serial Endosymbiosis Theory (SET), has steadily pushed it to centre stage in the biological sciences.

Put at its simplest, Margulis maintains that the principal source of evolutionary innovation, the origin of new species, is not only through the random mutation of genes edited by natural selection—the neo-Darwinist view—but through permanent merger of different strains of bacterial and other cells that descended from our common bacterial ancestors. Symbiogenesis she insists, is a source of evolutionary innovation.

Symbiosis may be defined as the living together of two or more organisms of different species in physical association for long periods of time. The sea anemone attaches itself to the shell of the hermit crab. The anemone provides the crab with camouflage, while stray bits of the crab's food nourish the anemone. At the level of an individual cell, permanent symbiosis can occur after one cell forces its way into another, followed by truce. Endosymbiosis refers to one type of organism inside another. Successive mergers—serial endosymbiosis—and joint survival, for example of both a swallowed cell and the swallower, have led to the formation of new species.

'My claim,' writes Margulis in her 1998 book *Symbiotic Planet*, 'is that, like all other apes, humans are not the work of God but of thousands of millions of years of interaction among highly responsive microbes. This view is unsettling to some . . . I find it fascinating.' Using endless evidence from her beloved microbial world, Margulis is out to show that living organisms not only compete and struggle, as the neo-Darwinians proclaim, they also interact and stick together.

The emphasis on symbiosis is not original to Margulis. It was a familiar notion among biologists earlier this century. But the association of germs with disease, the rise of neo-Darwinism and the triumph of molecular biology, with its focus on the 'selfish gene', marginalized symbiotic theory. When the young Lynn Sagan (Margulis's first husband was the planetary astronomer Carl Sagan) tried to publish her symbiotic theory of cell origins in the late 1960s, she met a brick wall—her paper was rejected 15 times and lost altogether by *Science*.

In it, Margulis proposed that the energy-producing organelles inside cells, known as mitochondria, were originally bacteria that had been swallowed by or invaded the cells; but she had no proof for this unorthodox idea. Then, in the 1970s, came clinching evidence. When mitochondrial DNA was analysed, it turned out to be different from the DNA in the nucleus of the cell and to be similar to that of oxygen-respiring bacteria. In other words, mitochondria proved to be naturalized immigrants, rather than native cellular citizens.

Today, the Margulis explanation of the origin of mitochondria is a staple of biology textbooks. But not in a form that satisfies its author. 'The exposition is dogmatic, misleading, not logically argued and often frankly incorrect. They say, this is "female" evolution. The "male" evolution is competitive Darwinism, and Serial Endosymbiosis Theory is cooperative, gentler, and kinder evolution.'

She feels that microbes have had a bad press, both among scientists and the general public. Most biologists, she says, dismiss microbes as primitive organisms: 'When Richard Dawkins says "lower organism", he means rat or bee. People generally don't know anything about microbes except that they've been told microbes are going to kill 'em. The terminology is military: microbes are labelled enemy agents, we have to conquer disease, etc.' She calls herself a 'microbial chauvinist' and is delighted by the fact that E. O. Wilson, who is a world expert on ants, concluded his autobiography *Naturalist* with a paean to microbes: 'If I could . . . relive my vision in the 21st century, I would be a microbial ecologist. Ten billion bacteria live in a gram of ordinary soil, a mere pinch held between thumb and forefinger. They represent thousands of species, almost none of which are known to science.'

As Margulis likes to point out, when Neil Armstrong stepped on to the moon and said 'One small step for a man, one giant leap for mankind,' he overlooked vast numbers of bacteria on his skin and in his intestine that stepped with him.

Her bug's-eye perspective has led to a passion to establish a different ruling taxonomy in biology, set out in *Five Kingdoms*, written with Karlene V. Schwartz. Most biologists, she says, are satisfied with a three-kingdom classification: plants, animals, and microbes. But this misrepresents evolution, because the ancestors of plants and animals were neither plants nor animals; and it leads to serious confusion in the classification of early organisms. Especially egregious is the term 'protozoa' (meaning 'first animals') to cover organisms, such as certain algae—which some still regard as plants. Margulis avoids 'protozoa' and prefers to speak of the much larger and more inclusive group 'protoctists', which means simply 'first beings'. The five kingdoms in order of their evolution are therefore: bacteria (cells without nuclei), protoctists (descendants of microbial communities that integrated and became cells with nuclei), animals, plants, and fungi.

The five kingdoms have important support, for instance from the veteran evolutionary biologist Ernst Mayr and from the palaeontologist Stephen Jay Gould. But there is a rival scheme, that of Carl Woese, proposing three kingdoms based not on an organism's structure or developmental biology but on the composition of its DNA. It is a case of biologists versus chemists, non-reductionists versus reductionists, says Margulis, who cannot resist a dig: 'Chemists often can't tell the difference between a live organism and a dead one.'

For almost her entire career, Margulis has battled on two fronts: for the symbiotic theory of cell evolution and for the Gaia hypothesis. She is still unclear how her two scientific passions are connected. ' "Mom, what does the Gaia idea have to do with your symbiotic theory?" asked my son Zach, aged 17, after work one day,' she begins her *Symbiotic Planet*. ' "Nothing," I immediately responded. "Or at least nothing as far as I'm aware." ' But she likes to quote a former student,

Gregory Hinkle, who quipped: 'Gaia is just symbiosis as seen from space.' She explains: 'If you get off the surface of the planet you see that organisms are in physical contact, through the surface waters and through the atmosphere, that is, organisms are always producing gases to the atmosphere or removing gases from the atmosphere. So from a distance, Gaia is simply symbiosis as seen from space.'

Even Margulis's strongest detractors grant that she is a fecund figure in today's biology. It may be that, to quote another former student Tom Wakeford, now lecturing at the University of East London: 'As so often in the history of science, Margulis herself has become so notorious that we have to wait for another generation to reap the full rewards of her insights.'

by Andrew Robinson

Proof Positive for Wet Mergers in the Eternal Takeover

LYNN MARGULIS

There are several scientific achievements in biology/geology that I anticipate in the twenty-first century. All of them relate to the central evolutionary question that Charles Darwin attempted to answer: what is the mechanism by which new forms of life are generated?

While Darwin clearly had part of the answer as to how some organisms survive and certain species leave descendants while others become extinct, he, and his successors, have failed to explain how new life forms arise in the first place. No one still believes that, by itself, the accumulation of random individual mutations in the genetic material DNA accounts for all of the evolution of the diversity of life on earth.

The achievements I anticipate will show that many of the fantastical claims about the origin of new forms of life made by ultra-Darwinists such as Richard Dawkins have explanations that are more complex and interesting than so-called 'selfish genes', mutating and reproducing under the conditions of a dead environment.

In the next twenty years, I suspect, we will see the definite explanation of the symbiotic origin of the first cell to have a nucleus. Symbiosis is the living together of two or more organisms of different species. At the level of an individual cell symbiosis can occur when a cell ingests, but fails to digest a bacterium. The bacterium divides inside the cell such that the establishment of a permanent partnership leads to a new life form, a new 'individual'.

We have already seen that the respiratory and photosynthetic apparatus of all modern cells came about via the intimate association of bacteria over a thousand million years ago. This merger process was dubbed symbiogenesis by Russian biologists a century ago. The final piece of the jigsaw is to show that the moving parts of a cell—the tails of swimming sperm cells, the rods and cones of our light-sensitive retina—also originated symbiogenetically from wary bacteria.

The twenty-first century will also see the end of the view that organisms can be fully characterized by the sequencing of their DNA alone, their genetic material. The belief that we could find the genes for complex cultural traits such as intelligence or environmentally induced diseases has been one of the biggest money-wasters of the twentieth century.

Another breakthrough of the twenty-first century will confirm what makes the cell the smallest living entity and will enable us to describe the cell's mimimal metabolic requirements. Instead of seeing an inert molecule like DNA as life's fundamental particle, we shall recognize that the wall-less bacterial cell, this tiny, self-forming entity, is in fact the 'atom', the basic unit of evolution.

Along with symbiogenesis, the mainstream of twentieth-century biology has ignored several other major mechanisms that explain the large discontinuities that have taken place in the evolution of new forms of life.

One of these is a large-jump chromosomal change that happens in mammals like carnivores and primates (monkeys and apes) called 'karyotypic fissioning'. Another is the idea of larval genetic transfer— the notion that different spineless marine animals can interbreed across the species barrier—as proposed by Don Williamson of Liverpool University. Sometimes transfers even seem to happen between whole different groups such as starfish and sea-squirts.

Further Reading

Margulis, L., *Symbiosis in Cell Evolution: Microbial Communities in the Archean and Proterozoic Eons*, 2nd edn. (New York, 1993).

—— and Sagan, D., *Slanted Truths: Essays on Gaia, Evolution and Symbiosis* (New York, 1997).

—— and Sagan, D., *What is Sex?* (A Peter N. Nevraumont book), (New York, 1997).

—— and Sagan, D., *What is Life?* (Los Angeles and Berkeley, 1999).

—— and Schwartz, K. V., *Five Kingdoms: An Illustrated Guide to the Phyla of Life on Earth*, 3rd edn. (New York, 1998).

Williamson, D. I., *Larvae and Evolution: Toward a New Zoology* (New York and London, 1992).

Don Norman

DON NORMAN CANNOT IGNORE A BADLY DESIGNED DOOR, TAP, OR LIGHT SWITCH. HE IS LIKELY TO PHOTOGRAPH IT, FIGURE OUT WHY IT IS UNFIT FOR HUMAN USE, AND PUT IT IN HIS NEXT BOOK. EVERYDAY LIFE MUST BE PRETTY ANNOYING FOR THIS MAN, ONE IMAGINES. ON THE OTHER HAND, EVENTS THAT MIGHT HAVE BEEN MAJOR FRUSTRATIONS IN HIS CAREER DO NOT SEEM TO IMPINGE ON THE CONTENTMENT THAT BEAMS FROM HIS BEARDED, BESPECTACLED FACE.

In 1983, after twenty-seven years in the psychology and cognitive science department at the University of California, San Diego, Professor Norman (he is still an emeritus professor at UCSD) decided academics were out of touch with the real world. Instead of agonizing over wasted years, he called some friends in industry, took the university's offer of early retirement and landed a job as an Apple Fellow, one of the Apple computer company's senior researchers, who pick their own projects. Explaining his decision to quit the academic life he quotes the management writer Tom Peters: 'If you are very happy in your job it is time to move on.'

With the decision to move out of universities Norman was deliberately opting for the real world. He certainly got it. When he joined Apple, the company was losing money. Shortly after Norman's appointment, a new management decided research was a dispensable luxury.

At such times, Norman has learnt, it helps to have a reputation. In the admittedly confined world of human–computer interaction, Norman has one of the best. His 1986 book *User-Centred System Design*, written with Steve Draper, 'defined the field of human computer interaction' according to Harold Thimbleby of Middlesex

University, himself a noted critic of ill-designed machines. And Norman's second, *The Psychology of Everyday Things*, was a best-seller.

So it was without too much effort that Norman found a welcome at another big computer company, Hewlett Packard. Once again the future looked bright. Everyone he initially met, including the company's research vice-president Joel Birnbaum, seemed to share his interest in information appliances—simple, specialized devices such as cellphones and palm-sized electronic organizers.

Unfortunately, it soon emerged that he had met the half-dozen information appliance enthusiasts in a company with more than 100 divisions. The ink-jet printer division was locked in battle with the laser printer division. Neither division was particularly receptive to wild new ideas from the labs.

Once again Norman got stuck in the role of design critic, confined to pointing out the advantages and disadvantages of different designs for particular machines. What he wanted at that time was to put his own ideas into action, get his own name on a successful product. He would have liked to lead Hewlett Packard into the age of the 'information appliance', building on the company's experience with calculators, scientific instruments, and gadgets that talk to each other by infra-red beam. Instead, hamstrung by the situation he found himself in, Norman wrote another book: *The Invisible Computer*, an attack on the design of most personal computers.

There is no gainsaying that, when it comes to exposing the design flaws in industrial products, Norman is peerless. Former Apple technology vice-president Larry Tesler says: 'Don is a superb and acerbic critic of machines and software that are confusing to use.' But some commentators argue that Norman is better as a design critic than as a designer. Harold Thimbleby, for instance, who observes: 'I think he is excellent at describing problems, but not at providing solutions.'

Norman rejects such comments. He points out that, even as a critic, he always strives to be constructive and often works with designers to solve drawing-board problems. He initially showed the draft of

The Psychology of Everyday Things (*POET*) to a group of industrial designers, who felt affronted by it and told him so over breakfast. So Norman rewrote the book. 'It is actually sympathetic to the design profession. I talk about the challenges they face and point out, for instance, how they don't have full control over the product.'

In the rewrite of *The Psychology of Everyday Things* what Norman gave back to the design community was a psychological explanation of why some designs are usable and some are not. 'It was one thing to rail about how badly things were designed. I also wanted to have a positive contribution, and try to point out what the principles really were.' He explains further: 'I sat down and asked myself the question, how is it that I get through everyday life without having had most of what I do explained to me? How do I know where the door is and how to open it? How do I understand everyday life? That was really the ultimate question.'

The answer Norman came up with drew on the work of the psychologist James J. Gibson: that every object has a set of 'affordances', in other words, every object has a set of uses to which it can readily be put. 'As I looked around it became clear that most of the information [about how I understand everyday life] was in the world. It did not have to be in the head,' says Norman.

Gibson was concerned with the practical affordances of natural objects. Norman had in mind something more artful, a set of perceptual cues that a designer could use to influence our behaviour. Doors are his favourite example. Some have handles which ask to be pulled, others have plates and bars which demand to be pushed. Only the designer who has failed to provide these cues has to resort to 'push' and 'pull' signs.

POET sold slowly at first. In paperback it was renamed *The Design of Everyday Things* so that bookstores would shelve it with books on design, rather than in the psychology section. Sales perked up. And with them, surely, the author's reputation? Norman sighs and explains again: 'You don't understand about the academic community. It was a popular book. Nothing is worse for your reputation than one of their members writing a popular book. Look at Richard Dawkins, you think he is respected for his popular books?'

But the hostility did not deter him. Once he had discovered his gift for popular writing he added *Turn Signals are the Facial Expressions of Automobiles* (1992) and *Things That Make Us Smart* (1993) to his credentials as the Dawkins of design.

In *The Invisible Computer*, his attack on that 1990s object of desire, the personal computer, was ferocious. 'Basically the PC is hopeless,' he told a London audience in 1998. 'There is no way to fix it.' It is too complicated, 'evil, even'. Something has clearly gone wrong when computer company helplines are blocked with calls from desperate users. Fridges and washing machines never caused such grief.

If anyone has the right to say 'don't blame me, I warned you', it is Don Norman. But he resists the urge to blame Microsoft's Bill Gates, or anyone else, arguing instead that all technologies—radio, the telephone, the phonograph—are fiendishly complicated in their early years and the PC is no exception.

The good news for the future, according to Norman, is that the work we used to do on PCs will be accomplished before too long on a variety of specialized information appliances, mostly small and without trailing wires. 'You will find you are using your PC less and less and at some point you will just stop using it,' he predicts.

The term 'information appliance' was coined in 1978 by Jef Raskin, originator of the Apple Macintosh project. But it has taken manufacturers years to create something as simple to understand and use as the US Robotics Palm III, a curvaceous steel clamshell that rests in your palm, understands your handwritten scrawl and quietly organizes your life. Norman pulls one out during our interview to look up an address, then gets drawn into the ritual of finding out who else he knows in Surrey. Putting it away eventually, he looks up apologetically. Even the best technology can be intrusive, he admits.

Future information appliances may be dedicated to domestic

finances or school homework. But the gadget most likely to give scholars pause for thought is a book-sized box with a screen on the front and enough chips inside to store a small electronic library. Two book-readers, the Softbook and the Rocket eBook, are now on the US market. And Norman has just visited a bookstore in Massachusetts which prints technical books on demand. 'They can print a book in 30 minutes or less, and then bind it. It looks like a medium-quality paperback book.'

For a machine to be successful, in Norman's view, not only must it be usable (his forte), but its engineering, marketing, and manufacturing must also be got right. 'It doesn't matter if you make the world's most usable products, if no one buys them. I want to make sure that people buy them, which means understanding what causes people to buy.'

In industry, Norman, usability's champion, met clever people whose concerns were different from his own. If he could not convert them, at least he could understand them. 'I believe I now understand the business model, the marketing model, the selling model, the design model, the usability model. Can't we put them together?' And this is what he is now trying to do.

He has left Hewlett Packard and joined forces with Jakob Nielsen, another top design guru who was encountering frustrations at Sun Microsystems. If they were rock stars, this would be a supergroup. The Nielsen Norman Group is the Emerson, Lake, and Palmer of human–computer interaction, and its supposedly quiet launch in 1998 was swiftly reported in top business magazines and newspapers—in *Business Week*, *Forbes*, and Silicon Valley's local paper, the *San Jose Mercury News*.

The idea behind their company is that Nielsen and Norman will take the design message to senior management in big companies, where it is rarely heard. Instead, things are unfolding in typical Silicon Valley fashion: 'We have been approached by several start-up companies, and they are paying us in stock,' says Norman. Big business has, so far, been slower to come to the bait. But there is plenty of time. Academia may have been unreal, and industry frustrating, but Don Norman's third career has only just begun.

by Tony Durham

Cognitive Prostheses

DONALD A. NORMAN

Technology changes rapidly: people change slowly. Evolution moves even more slowly: human beings have not evolved in significant ways throughout recorded history. But if biological, natural evolution moves slowly, technology moves quickly. Up to now, technology has acted as a supplement to human capability. A calculator does not enhance the brain, but a person plus a calculator is a far superior calculation device than either calculator or person alone. The calculator acts externally to the body: the power of the unaided mind has not changed. What will happen when the technology is small and powerful enough to be implanted within the body, perhaps directly connected to the brain: will technology supplement the mind?

Today we have artificial implants for bones, limbs, and organs. These are primarily mechanical in nature, although the pacemaker is an information device, helping the heart time its operations. We are starting to see sensory implants; artificial cochleas and retinas. I myself have two artificial lenses—plastic lenses instead of the normal biological ones. But what happens when the newest information technologies merge with biological technologies? What happens when we have implants within the body that affect our cognitive powers?

In the coming one hundred years we can expect major changes. Some technologies will start to encroach upon human biology and evolution. Some biometric alteration will occur. The future will see artificial sensory systems, communication devices embedded under the skin, perhaps computational systems as well to enhance memory, sensory abilities, physical strength, and language skills. Perhaps planning for a baby will be somewhat like ordering a new automobile: the proud parent(s) (singular? plural?) will be able to choose from a list of options: physical characteristics, hair and eye colour, bone structure, cognitive abilities. Will this planning and artificial implantation of technology be possible? Will it be permitted? If not, will bootleg sites spring up to do it anyway, perhaps in geographical locations that will thrive as legal havens for activities frowned upon in more traditional legal systems?

The basic needs of people are fixed by biology: the fundamentals have been with us for thousands of years and are unlikely to change within the next century. These include the necessity for food and shelter, for social relationships, for family bonding, education, sports, and entertainment. Sex, gender differences, emotions, and intellectual curiosity are likely to remain unchanged. Finally, people are motivated by a wide variety of drives, desires, and needs. Some of us work together harmoniously, cooperatively. Some compete and try to control. Some defraud, lie, and cheat. Cognitive prostheses must build on top of the existing biological substrate. Prostheses can enhance all these human aspects. But beware, the enhancement affects the positive and social as well as the negative and antisocial.

Consider a few types of potential prostheses: one for sensory systems (artificially enhanced seeing and hearing), one for memory, one for communication, one for emotional states.

Sensory. Built-in sound enhancers to make audition ever more acute, vision ever more precise. Built-in zoom lens. Built in recorder for sounds and sights. All of these are well within technological reach. It may be decades before they are small and reliable enough, but the ability to enhance the power and range of hearing and seeing is well within our grasp. Do we wish this to happen? Perhaps more relevant is the question of whether or not we could stop it if we wanted to: I suspect not.

Memory. I once wrote an essay entitled 'The Teddy' in which I predicted the development of a toy teddy bear that was given to a child at birth, a bear that could record the child's innermost thoughts, and aid in its development. As the child grew older, the teddy was replaced with more suitable devices—but preserving all the information from generation to generation, eventually being implanted in the brain, the better to allow the person to record all that had ever happened, all that was ever thought. I speculated upon the legal implications; would others be allowed to have access to the contents? What about with court orders? What about after a person's death? Who owned the recording?

Is this possible? Yes. Is it desirable? Not clear. Our fallible memories are blessings, for if we remembered too much, we would have trouble recalling the important items from amongst the trivia: thus the Russian neuroscientist Luria reported that one of his patients, a person who never forgot, found this to be a curse, not a blessing.

Communication. We already have pocket-sized cellular telephones. It won't be long before they are small enough to be embedded beneath the skin, perhaps in the flap just below the ear. Couple them to the nervous system and you have continual communication. How would society change if everyone could be in continual communication with others? Groups can be both more and less intelligent than individuals, a process much studied by social scientists. How would society change if one were never alone, if there were always someone talking to you, helping you, giving guidance, debating your actions. Or perhaps deriding your abilities, egging you on, falling prey to group decision making—consensus rather than bold imaginative steps, caught by emotions and rituals. The thought is not pleasant. The science-fiction author Vernor Vinge has speculated about a race of animals that were always in continual communication: their group intelligence was far superior to that of any individual. They had trouble fathoming human beings, for here, each 'singleton' had an intelligence approaching that of their group.

But why stop with memory enhancers and telephones? Language translation is just beginning to be possible. Within the next one hundred years it will certainly be practical, even if imperfect. What about implanted translators? Pitch synthesizers to enhance one's voice? Built-in calculators, maybe even a small memory chip of pre-coded information—encyclopaedias, reference works, and the like. Surely if we can embed a telephone we could embed a web browser, yielding its information as visual or acoustical images.

Emotions. Emotion is still the stepchild of the neural and cognitive sciences, less well studied and less well understood than the mechanisms of sensation and thought. Nonetheless, progress is being made in understanding the chemical, electrical, and anatomical substrates of emotion. What if implants would deliver precisely controlled squirts of neurochemicals, or small, controlled electrical signals to relevant brain centres? Pleasure centres could be enhanced, emotional feelings enhanced or suppressed. The devices would almost certainly start out as medical prescriptions for controlling malfunction, but could easily be subverted to enhance pleasure and to produce artificially enhanced mood states. Would we have a generation of brain zappers?

If these predictions seem outlandish and unsettling, they are mild compared to what others have proposed. I try to be restrained; others have suggested that we could do entire brain 'uploads', transplanting mind into electronic memories and devices. Thus, imagine the day when human wetware is replaced by software. Ugh. Fortunately, I do not believe this possible, certainly not within the timeframe imagined. Moreover, I do not think we will crack the internal coding of thoughts, emotions, and actions that would make direct communication with the nervous system possible; I think it will always be mediated through our existing sensory systems. I could be wrong. Scientists do not have a good record of prediction: things thought to be imminent can take centuries; things thought to be impossible can appear in decades.

What will happen in one hundred years? A lot. Which will come true, which will remain fanciful only time will tell: check back with me then. If medical technology progresses rapidly enough, I might still be here.

Further Reading

Norman, D. A., 'The Teddy', in *Turn Signals are the Facial Expressions of Automobiles* (Wokingham, 1993).
—— *Things That Make Us Smart* (Wokingham, 1993).
Kurzweil, R., *The Age of Intelligent Machines* (Cambridge, Mass. and London, 1990).
Solotaroff, L., Bruner, J., and Luria, A. R., *The Mind of a Mnemonist: A Little Book about a Vast Memory* (Cambridge, Mass. and London, 1987).
Stephenson, N., *The Diamond Age* (London, 1996).
Vinge, V., *A Fire upon the Deep* (London, 1992).

Paul Nurse

YOUR CELLS GOT YOU WHERE YOU ARE TODAY. ONCE THERE WAS JUST ONE. BY NOW, THERE ARE TRILLIONS OF THEM, EACH MAKING TWO NEW CELLS JUST OFTEN ENOUGH TO KEEP YOU GOING. BUT ANY ONE OF THEM COULD YET BE THE DEATH OF YOU. ANY ONE CAN BREAK RANKS, START REPRODUCING OUT OF TURN. SOME OF ITS DESCENDANTS COULD SHAKE OFF OTHER RESTRAINTS AND SEEK NUTRITIOUS SITES TO BUILD THEIR NEW OBJECT IN LIFE: A MALIGNANT TUMOUR.

This is why understanding cell behaviour is at the heart of cancer research. And why a man who, twenty-five years ago, set out to study cell division as a problem in pure science is today head of the Imperial Cancer Research Fund, the country's largest cancer charity. Paul Nurse's unravelling of his chosen problem shows the prescience, clear thinking, and luck that a successful scientist needs, in a combination that has made him one of the most celebrated biologists of his generation. Yet it nearly did not happen.

In the early 1970s, Nurse was half-way through a not particularly satisfactory Ph.D. at the University of East Anglia. He had been drawn to science since his schooldays, but the move from learning to doing was proving trying. 'As an undergraduate you read about the beautiful experiments that all work,' Nurse says. 'When you are a postgraduate and you do them yourself, they are not such beautiful experiments and most of the time they do not work.' Sitting up nights, keeping watch over a temperamental amino acid analyser, he wondered if the experimental life was really for him, and he toyed with the idea of giving it up for philosophy.

If not philosophy, what? He had turned away from his schoolboy interest in ecology and natural history—as a boy he plotted the distribution of birds and insects in different parts of his garden—in favour of the tightly controlled experiments of the biology lab. But he was still interested in whole organisms, while the molecule-by-molecule analysis that dominated the laboratory invariably meant ripping such organisms apart.

What might give him a way forward in science that combined a concern with systems that were still living with the control of the variables that he craved? And what was important enough to induce him to carry on in the face of all those failures, to deal with 'the cold water of having to do experimental manipulations all the time'? The answer to both questions was the same: the cell, the smallest piece of anything that is still alive.

Biology textbooks told of the life cycle of a single cell, which microscopic observation had recorded in great detail. It was a complex affair, with several different phases and wondrously ordered preparations for cell division involving chromosome copying, separation, and repackaging of the chromosomes into two cell nuclei where once only one existed. The diagrams depicting all this were labelled with many terms which taxed the memories of generations of schoolchildren, but the whole scheme was in essence descriptive: what might control when and how often a cell divides was unknown. Nurse decided to find out.

It must have required unusual self-confidence, though looking back now as an affable 50-year-old at the peak of his profession, Nurse makes it sound obvious. 'The problem in biology is how you make an organism, how organisms change in space and time. Seeing a cell develop was that in its simplest possible way, because a cell just grows, doubles its size, reproduces everything in itself, and divides. I decided that the cell cycle had all the characteristics of a problem that defined the interesting features of life.' Moreover, it should be open to study with the genetic techniques then coming into use.

Nurse had never worked on the cell cycle, nor did he know much about genetics. But he persuaded a more experienced researcher,

Murdoch Mitchison in Edinburgh, to take him on. He flashes a smile and says: 'I just got into the problem with a running jump.'

The best-studied cells were in yeast, and that was where he began. An American, Lee Hartwell, had paved the way with studies of budding yeast. Nurse would work with so-called fission yeast—the names indicate how the types reproduce. Before settling in Edinburgh, he spent half a year in Berne learning yeast genetics. Within a couple of years, he reported finding a yeast strain that split into two when it was only half the normal size for cell division. There followed a host of other mutant strains, about fifty were known by 1990, with alterations in their genes that affected the cell cycle in various ways.

The most useful strains were temperature-sensitive, permitting normal reproduction but getting stuck somewhere in the cycle if they warmed up. To find out what the products of all these genes do meant shifting from overall genetic mapping to the molecular level. After moving to Sussex University, Nurse set up a group to do this, building up libraries of normal yeast DNA and inserting pieces of DNA into mutant strains one at a time, to find out which one would kick-start a temperature-sensitive cell.

The most important early finding was a gene that woke up cells whose cycles halted after they made two sets of chromosomes but before they began to move apart. This gene enables the cell to make a protein that, when bound to another key player called cyclin, works on several other proteins to allow the cell to move on to division.

It was a stunning illustration of the power of genetic techniques to unravel complex phenomena. No wonder Nurse is irked by some portrayals of genetics. As he told a scientific meeting in London in the early 1990s: 'I'm a geneticist, and I get sick of the fact that we are always being pilloried as Neanderthal reductionists, one step away from a Nazi concentration camp warden. Genetics looks at the whole organism.'

Further work on the gene Nurse had found showed that this controller is controlled, in turn, by other proteins. The details rapidly become too complex to relate. Read the papers and you understand the force of Nurse's comment that one limit to a programme to enumerate every molecule in a cell is boredom. But the basic principle is simple. There is a fundamental mechanism of switches in cells—used to control progress through the cell cycle, among other things—that depends on a relatively simple chemical change. The switches themselves can be changed in the same way, and they can often sense the presence of other chemicals. Combining such elements means that you can build complex networks of signals whose components interact with each other and with other parts of the cell. A whole set of sensors and switches that automatically responds to the levels of particular molecules present at any particular time can coordinate the movement of one cell towards becoming two.

By now, the relevance of all this to cancer was clear, and Nurse moved to the Imperial Cancer Research Fund's laboratories in London's Lincoln's Inn Fields, leaving for a stint as professor at Oxford University before returning to the ICRF as director of research, then director-general. Now he has to help fashion scientific strategies for hundreds of scientists in dozens of laboratories, but he is still mindful that the work can grind you down and that the really satisfying moments are few and far between. The bottom line is that the problems must be worthwhile, he says, because if the stakes are high, 'it's worth the pain'.

No one could argue that the stakes are not high in cancer, which has just overtaken heart disease as Britain's number one killer. An organization with a mission to tackle cancer must work with treatments for patients and schemes for prevention. But Nurse the biologist is equally committed to longer term programmes—he speaks of part of the institute's portfolio being 'risk-invested'. His own career is a good example of the way such investments may pay off—we now know that the proteins that regulate the cell cycle in yeast do the same job in humans.

Nurse's work still focuses on the simpler organism. At the ICRF he has managed to do what most scientists-turned-administrators rarely achieve: keeping in touch with the lab. He says he spends half his time running his research team and half running the organization, but he admits that others' idea of half time in the lab may be different from his.

Aside from refining details of the cell cycle, his team is working on another problem—essentially, how cells know which way is the right way up. Already, they have found a protein that locates end-markers in yeast cells. Again, the possible connections with cancer are there to be explored, for cancer cells change shape when they migrate. So working out how a cell knows what shape it is in the first place should help explain such changes.

Together, the two investigations bracket properties of cells that Nurse emphasizes cannot be studied just one molecule at a time: organization in time and organization in space. If all of the control elements that regulate the cell cycle can be identified, for example, what will the list tell us? Even if their connections are drawn in a pretty diagram on paper, it is likely to be a pale reflection of what is actually going on inside the rich little world of the simplest living unit. Nurse wants to move beyond the naming of parts: 'The methodologies we are good at tend to lose information.' Understanding the cell cycle, he suggests, will mean 'actually describing, in real cells in real time and space, how these signalling pathways work'.

Suppose, for example, that what matters is not whether a particular signal is activated but how long it is activated. Suppose, in fact, that cells have some kind of internal Morse code. Then consider the distances messages written in such a code must travel, which are small by our standards but very large compared with the size of a protein molecule. So how can cells monitor their internal states as three-dimensional empires with outlying provinces of proteins as well as nucleic acid heartlands?

'We may not find it easy to think about these sorts of problems,' Nurse reckons, but he is looking to control techniques to produce new ways of describing the emergent complexities of the cellular interior. In the end, though, however much this language speaks to the limits of one kind of reductionism, practical approaches to cancer will demand going down to the next level. As he says, 'when you are interested in engineering and manipulating a system, you have to go down to the molecular level because that is the only way you are going to be able to operate on it.'

He still thinks that major improvements in cancer treatment—the hope that secures the £1 million a week the ICRF needs to keep going—are thirty years away. That could at least mean that the scientists who deliver such improvements have already started work, even if they are still at the stage Paul Nurse was when he first fixed on the cell cycle as the part of life he really wanted to understand. And when they come, the treatments will almost certainly have one thing in common; they will eliminate the cells that cut loose from the controls on cell division that Nurse helped to make visible.

by Jon Turney

Complex Cancers, Simple Cells

PAUL NURSE

Living things are extraordinary objects—complex self-organizing systems which can grow, develop, and reproduce. The basic unit of life is the cell, the simplest entity exhibiting the characteristics of life. So, the breakthrough I would like to see in the twenty-first century is a complete understanding of the workings of a cell, which in turn will provide real insights into the nature of life.

At this point in time biologists are in the position of being able to count, name, and describe all the genes found in the simplest cells. The real problem for the future will be to work out how the products of all these genes act together and operate within the cell so it acts in certain ways. Two particular problems need to be solved. The first is the ability to follow the chemical reactions within the living cell in real time and space. Living cells have a whole set of sensors, switches, and timing devices, which must all be based on interactions between the molecules inside them. Traditionally these have been studied outside the living cell 'in vitro', that is, within the glass test tube.

Advances in chemistry, biophysics, and microscopic imaging will make it possible to analyse these reactions within a single living cell. We will be able to trace events in cellular time and space, and describe properly the continual changes inside cells in terms of the molecules involved.

The second problem is to be able to study how all the molecular components of a cell are linked together in complex networks. When molecules interact in particular ways, networks emerge which underpin the special ways in which living cells behave. A simple example is the 'negative feedback' network, which senses some process has speeded up, and slows it down again—rather like the governor of a steam engine. We may find molecular examples of more complex devices, like oscillators which help regulate cycles of cellular events in time, or ways of controlling the diffusion of molecules which offer spatial regulation.

Central to this thinking is information flow and processing. It may be necessary to represent the molecular networks of a cell not only through bottom-up numerical modelling, but also through symbolic depictions analogous to electronic circuits.

Combining these two approaches will provide new understanding into the nature of life. It represents a major intellectual challenge, tackling the very question of what is life. It also provides new approaches for understanding many diseases, some of them immensely complex such as cancer. It is only through new understanding that new and better treatments will be developed.

Further Reading

Bock, G. R., and Goode, J. A., eds., *The Limits of Reductionism in Biology*, 1998 Novartis Foundation Symposium 213 (Chichester and New York, 1998).

Coen, Enrico, *The Art of Genes* (Oxford and New York, 1999).

Mayr, Ernst, *This is Biology* (Cambridge, Mass., and London 1997).

Smith, John Maynard, and Szathmáry, Eörs, *The Major Transitions in Evolution* (Oxford, New York, Heidelberg, 1995).

Webster, Gerry, and Goodwin, Brian, *Form and Transformation* (Cambridge, New York, Melbourne, 1996).

Steven Pinker

WHEN STEVEN PINKER, EAGER TO INVESTIGATE THE WORKINGS OF LANGUAGE AT FIRST HAND, DECIDED TO USE REAL CHILDREN IN HIS RESEARCH, HIS ACADEMIC ADVISER ISSUED A WARNING: 'YOU'LL NEVER REALLY UNDERSTAND WHAT'S GOING ON IN THEIR MINDS.'

Undeterred, Pinker has dedicated himself to finding out what is going on in all our minds—not just in terms of language but in terms of everything else too.

So his best-selling book *The Language Instinct*, which explored how language develops, was followed a couple of years later by *How the Mind Works*, a lengthy paperback which starts with a chapter on the mind's 'standard equipment' and ends with 'the meaning of life,' negotiating examples from the television show *LA Law*, Gary Larson cartoons, and *Indiana Jones* along the way as 'little rewards for the reader'.

These two books and the academic rows which followed them have propelled Pinker into the big league of international scientists.

He carries his academic credentials—professor of psychology and director of the Centre for Cognitive Neuroscience at the Massachusetts Institute of Technology—alongside the style of a 1970s rock star, with a mane of well-kept curls framing a chiselled jaw and a penchant for designer suits and coloured ties. Fame, he says, 'is kind of fun'. Occasionally he is recognized in the street but people do not always know who he is. Sometimes, he says, with a slight smile and tug of that hair, they think he is the conductor Simon Rattle.

The only time celebrity has worried him is when he wrote an article suggesting that women who murdered their new born babies were not necessarily mentally ill. Rather, he proposed, they might be unconsciously obeying primitive instincts. 'Conventional explanations [for infanticide], such as childhood abuse [of the mother] or insidious rock lyrics, were unlikely,' he says. 'More likely was a selective engagement or non-engagement of maternal instincts depending on circumstances.' This view, which he knew would be controversial, was targetted for attack by the American Christian right-wing. 'I got flooded by e-mails—some of them threatening,' he says. Painstakingly, he replied to every communication—'I didn't want them to think I believed in infanticide'—and the fuss eventually died down.

Pinker is endlessly patient when there is something to be explained, leaning forward and spooning out phrases, heaped with images, in a soft voice. His explanation of how the mind works is as 'a system of organs of computation, designed by natural selection to solve the kinds of problems our [Stone Age] ancestors faced in their foraging way of life, in particular, understanding and outmanoeuvring objects, animals, plants, and other people.'

The mind is not the brain but 'what the brain does' and what the brain does is process information. To do this, it is organized into 'mental organs', each designed by a genetic programme and shaped by natural selection to perform a specific function. Unravelling it all demands a kind of 'reverse engineering', identifying what each 'mental organ' is designed to do and then working out how it has developed to do it.

While Pinker suggests that 'our minds lack the equipment to solve the major problems of philosophy', his view of the world offers an answer for almost everything else. Children in cities fear snakes because, when humans were hunter-gatherers, they needed to fear them. People like gossip because knowing what other people do not know gives strategic advantages in the games of life. They appreciate art because then they 'gain status through cultural machismo or artificially stimulate their pleasure circuits'.

And language—the gift humans alone possess—is just one of the facilities we have to deal with the world, evolved over thousands of generations to enable better communication and hence increase our chances of surviving long enough to have children and pass our genes on to the next generation. 'Anthropologists have noted that tribal chiefs are often both gifted orators and highly polygynous,' writes Pinker. [This] 'is a splendid prod to an imagination that cannot conceive of how linguistic skills could make a Darwinian difference.'

This belief, that the development of human language is the direct product of natural selection, that language developed because it gives people an advantage in the reproductive marketplace, is one way in which Pinker's work diverges from that of the American linguistics professor Noam Chomsky.

Way back in the 1950s Chomsky came up with the then highly controversial theory of a 'universal grammar', arguing that every child is born with a sort of mental template for grammar which enables them to create sentences they have never heard before. Chomsky, before Pinker, thought that linguistic ability is innate. Where they differ is over how and why language developed.

Pinker's view is that language is an 'adaptation', a direct product of natural selection, and that it has been 'selected' over generations to satisfy the function of better communication. Chomsky disagrees, saying that there is, as yet, insufficient scientific evidence to support such speculation. His view is that 'the question of the evolution of the brain is far beyond the reach of the theory of evolutionary development.'

Chomsky complains that, in *The Language Instinct*, Pinker, exhibits 'confusions about evolution' and that one common confusion is to assume that natural selection is the single factor in evolution rather than one factor among many. 'Possibly,' says Chomsky, 'he [Pinker] shares the common belief that Darwin tried so hard to counter: that natural selection is evolution. Of course that could not be. In fact the proposal cannot be coherently formulated. Natural selection takes place under conditions of physical, chemical, biological law, including principles . . . involving complex systems and the way they develop—principles about which very little is known.'

But, says Pinker, 'it is Chomsky who is confused'. He points out that in his book *The Language Instinct*, he specifically writes 'Evolution and natural selection are not the same thing'—and goes on to explain why.

Pinker has conducted experiments with children to explore how language works. When asked to compare his work to Chomsky's, he says they are complementary but different. Though inspired by the work of Chomsky and other linguists, he says, 'I am interested in the nitty gritty of language use—how you account for real child language, for example, when, how, and why a child says "he breaked it".' Chomsky says 'looks more at the output—the logic of language in an adult'.

Pinker has taught new words to toddlers and observed how they use them in sentences. He has measured with computers the speed at which adults manage to convert a verb into the past tense. And he has focused on the way children use regular and irregular verbs and inflexion, seeing it as a way of encapsulating the two main processes of language: combining the memory of words already heard with the rules of mental grammar to make new sentences. This is to form the basis of his next book: *Words and Rules: The Ingredients of Language*.

Nonetheless, one of the most exciting aspects of *The Language Instinct* is his speculation that one day a 'a suite of genes' for language may be discovered.

Pinker's own childhood was spent in Montreal, Canada. His grandparents were Jewish Polish and Romanian immigrants—'it was a culture where there was a lot of arguing'—his father a travelling salesman and a lawyer, his mother a homemaker and then a guidance counsellor and high school vice-principal.

He studied psychology at McGill and, later, at Harvard, arriving in the early 1970s, in time for discussions about what human nature was about, but too late for the heyday of student radicalism.

Then he discovered the discipline of experimental psychology, which combined study of theories with study in a laboratory. His thesis was on visual imagery and spatial cognition and half his research was in that area until he decided to concentrate on language. This was not through any particular aptitude as a polyglot, although he speaks some French and a smattering of Spanish and Hebrew.

Rather, he discovered that the visual field was already crowded. 'Also, when I was 17, I read about Chomsky and it seemed tremendously exciting,' he says. 'I'm interested in big ideas.' But he never studied with Chomsky, although he did attend one of his courses. Nor does he see much of him these days, despite working in the same institution for the past seventeen years.

As director of the McDonnell-Pew Centre for Cognitive Neuroscience at MIT, where he moved after teaching at Harvard and Stanford, he still teaches a class of 300 students. Explaining his ideas to students helps him with his books for the general reader. 'You have to put yourself in the mind of someone who is intelligent and serious but who doesn't know what you know,' he says. 'You have to dismantle [your knowledge] and put it back together for someone.'

These books, which so far number four in total (the first two for a specialist readership), take application. He works at them seven days a week, often until 3 a.m. But he enjoys the popular works because he can write less defensively than for a scholarly audience. 'Academics tend to write everything in such a way that they cannot be nailed by a quotation,' he says. The problem is, by writing general books he risks people coming back at him in academic debate later.

And people do come back at Pinker. Although he has pages of clippings from reviews of his two last books praising them for their style and clarity—'a model of scientific writing: erudite, witty and clear,' says Steve Jones of *How the Mind Works*; 'his writerly charm helps to neutralise the acid that is Darwinism,' confirms John Horgan—there have also been some hostile reviews, some from academics reluctant to accept the grander claims of evolutionary theory.

Pinker says much of the hostile reaction to using evolutionary theory to explain psychological questions is because people think of adaptation in terms of what we *ought* to do. For example, when he uses evolutionary theory to explain tendencies among males to compete aggressively for dominance, people get angry. 'But there is a confusion between is and ought,' he says. The implication is that if we know why men behave like this—we can perhaps try to change their behaviour.

Another common confusion is to interpret theories of the selfish gene', driven to replicate itself, as producing selfish people. But sometimes, he says, the most selfish thing a gene can do is to produce a selfless person.

Pinker, 44, and his second wife Ilavenil Subbiah do not have children. 'Well into my procreating years I am, so far, voluntarily childless, having squandered my biological resources reading and writing, doing research, helping out friends and students and jogging in circles, ignoring the solemn imperative to spread my genes,' he writes in *How the Mind Works*, '. . . and if my genes don't like it, they can go jump in the lake.'

Far from finding it ironic that an evolutionary theorist is voluntarily childless, he uses it 'as an example of the point that evolution shapes emotions to be adaptive on average in an ancestral environment and does not shape individual people's behaviour in modern environments.'

He and his wife may choose to have children at some point in the future, he says. But not yet. For now, his contact with infants will be through his work, as subjects of his language research. Again comes the slight smile as Pinker recalls another warning from a colleague. Apparently, the man, a new father, found his pet theory in tatters as soon as his own child spoke.

by Harriet Swain

Increasing Consilience

STEVEN PINKER

I predict that the major development in the study of mind in the next decades will be an increasing 'consilience', as E. O. Wilson calls it, among the branches of human knowledge. Just as earlier centuries had to obliterate the once-absolute and now-forgotten divisions between the past and the present, and the living and the non-living, so we will obliterate the distinction between biology and culture, nature versus society, matter versus mind, and the natural sciences versus the arts, humanities, and social sciences.

New disciplines at the frontiers of biology, in particular, behavioural genetics, evolutionary psychology, and cognitive neuroscience, are laying a bridge between nature and society in the form of a scientific understanding of human nature. Our genetic programme grows a brain endowed with emotions and with learning abilities that were favoured by natural selection. The arts, humanities, and social sciences are about the products of faculties of the human brain: as language, reasoning, a moral sense, love, an obsession with themes of life and death, and many others. As human beings share their discoveries and allow them to accumulate over time, and as they institute conventions and rules to coordinate their desires, the phenomena we call 'culture' arise.

A fundamental division between the humanities and sciences may become as obsolete as the division between the celestial and terrestrial spheres.

Concretely, I expect that the completion of the human genome project [the project to unravel mankind's genetic blueprint] will lead to a sudden jump in our knowledge about the genetic basis of our emotions and our learning abilities. (The recent discoveries of sets of genes associated with speech, spatial cognition, general intelligence, and several personality traits are a forerunner of far more extensive and sophisticated analyses of the genetic basis of mind.)

In evolutionary psychology, a combination of mathematical and computer modelling with data from ethnography and experimental psychology will lead to an elucidation of the adaptive basis of major mental abilities. Recently we have gained tremendous insight into mysteries such as beauty, violence, sexuality, reasoning, and family conflict from evolutionary psychology, and I expect that the theory of natural selection will expand its importance as a constraint on theories in psychology and neuristic experimentation. And most obviously, the current explosion of research in cognitive neuroscience (the study of the neural basis of cognition, and increasingly, of emotion), will continue.

Techniques such as functional Magnetic Resonance Imaging, Magnetoencephalography, and Optical Imaging will become the standard empirical tools of psychology.

I personally find this all to be an exhilarating prospect, but I know that it is not an innocuous one. Scholars in the humanities often see biologists as carpetbaggers with little feel for the richness of their subject matter. Other people see the explanation of mind in physical terms as an undermining of meaning, dignity, and personal responsibility, or misread claims that certain motives are 'typical' of humans as being claims that they are 'justifiable'. Therefore I expect that all of these fields of study, and particularly those tied to genetics and evolutionary biology, will draw heavy, sometimes vituperative, criticism. The criticisms can and must be addressed, and I suspect that in the long term, connections between the study of mind and society and the study of genes, brains, and evolution will become solid and widely accepted.

Further Reading

Pinker, Steven, *The Language Instinct* (New York, 1994).
—— *How the Mind Works* (London, 1998).

Sherwood Rowland

WHATEVER HAPPENED TO THE CHEMICALS KNOWN AS CHLOROFLUOROCARBONS OR CFCS—ONCE THE BASIS OF A $200 BILLION AEROSOL SPRAY INDUSTRY?

CFCs were created in the 1920s when chemists working for the giant Du Pont corporation came up with a new propellant—a mixture of two gases, compounds of chlorine, fluorine and carbon—that could be stored in small metal tins. When, in the 1950s, a valve was invented that could squirt the can's contents at the touch of a finger, the stage was set for the emergence of what became, albeit briefly, a hugely lucrative product.

Up until a few years ago, aerosol sprays containing CFCs were still in use, wafting forth everything from insecticides to deodorants. In the 1970s came the first hint that the gases might be inexorably destroying the earth's atmosphere, threatening to kill us all.

The United States was the first to act. It banned aerosol propellants in 1976. Canada and Scandinavia followed suit. But the world didn't sit up and take notice for another decade—until, in the mid-1980s, scientists discovered a potentially catastrophic hole in the ozone layer that shields the earth from the sun's cancer-causing ultraviolet rays. In 1996, thanks to a worldwide ban, CFC production ceased.

The tale of how the world went from news of a problem to a total ban within two decades is also the story of the career of Professor Sherwood 'Sherry' Rowland. For first alerting the world, in June 1974, to the fact that CFCs were destroying the ozone layer, Rowland shared the 1995 Nobel prize for chemistry with his former postdoctoral researcher Mario Molina and a Dutch collaborator, Paul Crutzen.

Since the 1970s this remarkably tall, calm, organized man, now 72, has spent most of his working life dealing with the fallout from that crucial discovery. 'It just sort of exploded,' he recalls. Although, like many scientists, Rowland is reluctant to engage with politics, he resolved, in the face of industry opposition to his findings, that it was his duty 'to make the scientific facts clear'. At first it was a question of 'deciding whether to pick up the phone to people like you,' he says, bluntly. 'I had to think about whether that was part of what a scientist should do.' Once he started being quoted regularly by the media, invitations to address senators and advise governments came flooding in.

Rowland is good at explaining science to the layperson—his descriptions of complex chemical reactions are clear, measured, and logical. Yet it can't have been easy to abandon the relative anonymity of research for such a high-profile role. Rowland was vilified and his research ridiculed; even his Nobel ceremony in Sweden was picketed. But at the end of the day his arguments prevailed. One US magazine, possibly exaggerating only slightly, dubbed him 'the man who saved the world'.

He is still keen not to be tied too closely to any one political party (though he admits to being an admirer of US vice-president Al Gore, one of the most knowledgeable politicians about environmental science). Indeed, as he is quick to point out, his first 20 years of research were paid for by the Atomic Energy Commission, scarcely the most environmentally-friendly organization. In fact his 1953 discovery that the amount of radioactive hydrogen in the atmosphere had increased over a three-year period languished unpublished for years because it pointed to the then secret information that the USA was making hydrogen bombs and that there had been a leak at one of the country's plants.

Further evidence of Rowland's scientific impartiality can be found in early research that infuriated environmental campaigners. A comparison of the amount of mercury in tuna fish long dead to that contained in recent kills revealed no significant increase. As he explains, the research showed that: 'maybe mercury in tuna is hazardous, but not because of something mankind has done.' It was a blow to those eager for hard evidence to support their campaigns to outlaw the dumping of waste and sewage in American lakes and rivers.

In fact, Rowland didn't start out with a direct interest in environmental research. His early work was on radioactivity, first at the Brookhaven National Laboratory and then, later, at the University of Kansas. In 1964 he was hired to set up a chemistry department at the newly opened Irvine campus of the University of California, where he still runs a research group. 'Then it had 1,500 students, now it has 20,000,' he says. But by 1970 he was getting restless. 'I was taught that if you wanted to carry on doing interesting work, then every so often you had to tear yourself away from what you were doing and start all over again,' he explains.

1970 saw the first Earth Day parade in the USA. Rowland had teenage children and his wife was interested in environmental issues. Casting around for a new research problem he hit first upon the fish. Then he heard that the British scientist James Lovelock had detected CFCs in the atmosphere, though Lovelock initially regarded their presence as harmless.

The chemicals had been specially chosen by the Du Pont scientists to be inert, to react with nothing. 'But as chemists we knew that they would react eventually,' recalls Rowland. 'I wondered what would happen to them.' He and Molina knew that the CFCs would eventually drift up to the upper stratosphere, ten or fifteen miles above the earth's surface. They discovered that, at that height, the sun's intense ultraviolet light would break down the compounds, forcing them to release their chlorine atoms and initiating a chain reaction. 'The significance of this chain reaction was that one chlorine atom could end up destroying 100,000 ozone atoms. If you multiplied that by the million tonnes of CFCs released each year, it soon became obvious that there was a serious problem here,' says Rowland. 'We realized within two and half months that what had been an interesting intellectual exercise was a major environmental problem.'

One night in the early 1970s, Joan, his wife asked him how the work was going. Rowland replied in a matter-of-fact tone, 'very, very well—but it looks like the end of the world.' His wife responded by throwing out all fifteen of the aerosol cans in the house. 'My thought then was fifteen down, six thousand million to go,' Rowland recalls.

The next fifteen years were a roller coaster of emotions, as governments and scientific committees met to discuss the problem, pulled one way by fear of a global crisis and the other by the multi-billion dollar industries that relied on CFCs. 'The idea that spraying underarm deodorant could contribute to a global environmental problem was seen by the business community as environmentalism gone mad,' says Rowland.

Despite this attitude, things seemed to be going well when the United States, Canada, Sweden and Norway all agreed to ban aerosol sprays containing CFCs as early as 1978. But that early progress soon slowed and no new bans emerged for another decade.

A powerful industry lobby and a new government led by Ronald Reagan proved determined to dismantle environmental legislation. As atmospheric scientists tried desperately to refine their experimental models, their computers creaking under the strain, predictions of ozone losses went up and down on a roller coaster of their own. Conflicting predictions were seized on by spokesmen for the CFC industry as proof of the inadequacies of this environmental science.

As more details of the hundred or so chemical steps involved in breaking down CFCs were discovered, and bigger and better computers became available for modelling the atmosphere, something happened that no scientist could have predicted. In 1985, the Cambridge-based British Antarctic Survey reported a massive 'hole' in the ozone layer above the Antarctic. For part of the year—September/October, the Antarctic spring—ozone levels were 60 per cent lower than they had been previously.

The world was shocked into action. In 1987 the United Nations drew up the Montreal Protocol calling for a 50 per cent reduction in CFC emissions by 1999. More evidence gathered implicating CFCs and that quickly became a total ban. In 1992 the deadline for the time by which the world would cease to allow production of these chemicals altogether was brought forward—from 1999 to 1 January 1996.

'If you had asked me back in December 1973, would we have an international agreement to ban CFCs in 13 years and a total ban in 17 years, I would have said it was not very likely,' says Rowland. 'When we started out we said, "this is what will happen in the future". But suddenly we were talking about a 50 per cent [ozone] loss in Antarctica today. It's the timescale that drove people to action.'

Since the implementation of the Montreal Protocol and the discovery of new compounds that work nearly as well as CFCs without damaging the atmosphere, the amount of CFCs in the atmosphere has stopped rising. But the damage already done is irreversible and will be measured in higher rates of cancer. Rowland calculates that the ozone layer has already been depleted by 10 per cent worldwide and another five per cent loss is likely. Yet he is optimistic. The Montreal Protocol, he thinks, is the forerunner of future international agreements that will need to be devised in response to new environmental threats.

Global warming, caused by rising levels of carbon dioxide in the atmosphere as a result of burning fossil fuels like oil, gas, and coal, is now the big issue on the agenda. The action taken so far to try to limit carbon-dioxide emissions has been nothing like as effective as the moves to ban CFCs. What is needed, according to Rowland, is another stark warning from nature. 'The question now in talking about climate change is what is the geophysical or biological equivalent of an Antarctic ozone hole, something so stark that it really makes people sit up and take notice,' he says.

What kind of climate change could be so striking? No one knows but possibilities might include the shutdown of the Gulf Stream, the large current of warm water flowing north-east across the Atlantic, or a permanent state of El Nino, the devastating climatic effects caused by changing sea currents off South America. The last time that the Gulf Stream failed, around 11,000 years ago, Northern Europe was plunged into a mini ice-age. The effect of a permanent El Nino would be no less serious. But while he is careful to say that these are not predictions, he points out that they are not outlandish suggestions either.

'After all what's the use of having developed a science well enough to make predictions, if in the end all we're willing to do is stand around and wait for them to come true?' he asks. 'The question is: can we act before rather than after some really significant change has happened that is either irreversible, or reversible only on a timescale of many centuries?'

Since 1978 he and his team, in research funded by the American space agency, NASA, have been taking air samples from across the globe, building up 'a fingerprint of the atmosphere from Alaska to New Zealand'. They have monitored smog in the South Atlantic which drifted from burning rain forests in Africa and, in an aircraft five miles up near Fiji, encountered pollution which 'if you were in Los Angeles would have violated Environmental Protection Agency standards'. The idea is to publish their findings—'then governments can't hide the levels of pollution they are permitting.' As Rowland says, the idea of containing pollution is long gone; now it spreads across continents and oceans.

Once again he is facing opposition from big business, worried about the cost of implementing measures designed to trap rather than release the surplus carbon dioxide produced by factories and power plants. But Rowland says that the cost of not taking action is never added to the equation. Global warming could cause sea levels to rise so high that South Florida would be submerged within a century. 'No one I know from the oil companies is putting that cost in the scales. We need to think about how quickly we want to have the climate of the earth changing.'

While public concern about climate change is growing, at least in the USA, he worries that there is a lot of misinformation. 'There is powerful lobbying from the fossil fuel industries and the groups that they fund. Our polls tend to show that public belief in the reality of global warming is growing, but probably not to the extent that the public are willing to do very much about it.'

All but a handful of climate scientists believe that global warming is already upon us and that sceptics have been given too much airtime. 'There has been remarkably little talk of the fact that every month of 1998 was the warmest since reliable records began. June of 1997 was warmer than any June for 125 years and this record-breaking continued through 1998 for eighteen months in a row. What would you like to convince you that we are experiencing global warming?' asks Rowland. 'If you start flipping a "warmer vs. cooler" coin and it comes up "warmer" eighteen times in a row, you start to suspect that maybe both sides of the coin say "warmer" and it is no longer chance.'

by Sian Griffiths and Ayala Ochert

Sequestration

F. SHERWOOD ROWLAND

During the 1980s humanity discovered—with the appearance of the Antarctic ozone hole—that its everyday activities could change the global atmosphere in a near-catastrophic manner. For most people this was a great surprise because of the giant disparity between the continent-size measured ozone loss and the tiny scale of the individual activities contributing to that loss.

Two multiplying forces are in operation here, and their combination furnishes an important warning for the condition of Earth's atmosphere as we enter the twenty-first century. First, the major processes whereby ozone is lost from the atmosphere involve chemical chain reactions by which each chlorine atom released in the stratosphere can remove 100,000 molecules of ozone before eventually falling back to earth's surface as hydrogen chloride. The second multiplier is the rapidly growing population of the Earth which reaches 6,000,000,000 this year, nearly four times the 1,600,000,000 who were alive in the year 1900.

In 1999, 85 per cent of all industrial energy came from our burning of the fossil fuels, coal, natural gas, and oil. Nuclear and hydro power provided most of the rest, with wind, solar and other renewable energy sources down at the 1 per cent level. The central atmospheric problem associated with the burning of these fossil fuels is the release to the atmosphere of the combustion product carbon dioxide, with the global warming and climate change which are its consequences. Or, to state the problem more precisely, it is the failure to prevent the release of carbon dioxide to the atmosphere after burning has made available the desired energy.

Another atmospheric pollution by-product of burning fossil fuels, especially in cities, is the simultaneous release of nitrogen oxides and still burnable carbon compounds—carbon monoxide, ethylene, etc.—leading to the formation of tropospheric ozone. Tropospheric ozone is one of the chief ingredients in smog—a huge pollution problem in almost every major city in the world. However, not all the growing global burden of tropospheric ozone is caused by traffic— much the same chemistry accompanies the burning of forests and waste farmland. Unfortunately, the likelihood is that the global prevalence of smog will rise in the next century because more and more people will use cars.

Many of the improvements in quality of life now enjoyed by the world's more affluent inhabitants are often described as 'labour-saving', which indeed they are. But it is not just that electric labour has been substituted for hand labour, but that there has been a multiplication in the amount of labour performed. When labour is no longer limited by having to be done manually there is a huge increase in the amount of work each person can do—and in the amount of energy expended.

In the next few decades, the increase in global population is a certainty. By 2050, the present 6,000,000,000 people in the world will very probably have expanded to 9,000,000,000. However, the demographic trends have slowed the yearly increase in global population from its maximum of about 100,000,000 per year in the 1980s, and the possibility exists that the total world inhabitants in the year 2100 may be no larger, perhaps even less, than in 2050.

Nevertheless, all of these extra people plus those of the present population not currently possessing a high quality of life are going to aspire to a higher standard of living. Future improvements will come, as they have in the past, by supplementing and replacing manual and animal labour with inanimate, much larger energy sources. And, for the next few decades, the major energy sources are highly likely to involve the combustion of coal, oil, and natural gas.

The twenty-first century will therefore begin with three major atmospheric problems firmly entrenched on a global basis: stratospheric ozone depletion, the greenhouse effect from increasing carbon dioxide and other trace gases—with accompanying global warming—and urban and regional smog. The 1987 Montreal Protocol of the United Nations, as later modified, has brought a worldwide ban on the primary chemical causative agents of ozone depletion, including the main culprits, CFCs. Measurements in the global atmosphere now confirm that the further release of CFCs has almost ceased, and bulk manufacture has stopped on a global basis.

In the most affluent cities, stringent controls on gaseous emissions have made progress against smog, but globally the problem is worsening. Here, at least, the path towards control has been set, but action is lacking.

The global increase of temperature, especially in the 1990s, culminating in record-breaking 1998, is drawing more and more attention to the question of climate change, and to the problem of how to control carbon-dioxide emissions. The Kyoto Protocol for handling the carbon-dioxide problem is modelled after the Montreal Protocol, but has much more modest goals in the face of a much larger problem. Kyoto essentially calls for stabilization of carbon-dioxide emissions, or relatively small cutbacks, while the Montreal Protocol prescribed an absolute ban on emissions of CFCs.

My expectation in the coming several decades is that the climatic consequences from continued greenhouse gas emissions will be more and more noticeable, and much more ominous. It seems probable to me that attempts to manage the carbon-dioxide problem will move from a focus on limiting the combustion of fossil fuels to attempts to prevent the release to the atmosphere of carbon dioxide after combustion.

'Sequestration' is the descriptive word for attempts to trap carbon dioxide after combustion and place it in some secure place. A Norwegian company, drilling in the North Sea, for example, has found a natural gas source under the sea bed, with carbon dioxide mixed in with the methane. They separated the two and put the carbon dioxide back under the sea floor.

This problem is multifaceted, expensive, and far from a solution. I think energy production in the year 2100 will continue to depend on fossil fuels, but that sequestration of carbon dioxide will probably be required by international regulation. By the end of the next century I think sequestration of carbon dioxide emissions will be a near universal industrial practice.

Whether serious climatic change will have been forestalled by these changes will be a continuing scientific concern throughout the next century and that will depend upon properties of the climate which are currently not fully understood.

Further Reading

Rowland, F. Sherwood, and Molina, Mario, 'Ozone Depletion: 20 Years after the Alarm', *Chemistry and Engineering News* (1994), 72, 8–13.

Rowland, F. Sherwood, 'Stratospheric Ozone Depletion by Chlorofluorocarbons (Nobel Lecture)', in *Les Prix Nobel* (1995), The Royal Swedish Academy of Sciences, 1996, pp. 207–30, also in *Angewandte Chemie* (1996), 35, 1786–1798.

Blake, Donald R., and Rowland, F. Sherwood, 'Urban Leakage of Liquefied Petroleum Gas and Its Impact on Mexico City Air Quality', *Science* (1995), 269, 953–6.

Dotto, Lydia, and Schiff, Harold, *The Ozone War* (New York, 1978).

Nance, John., *What Goes Up: The Global Assault on Our Atmosphere* (New York, 1991).

Roan, Sharon, *Ozone Crisis: The 15 Year Evolution of a Sudden Global Emergency* (New York, 1989).

Amartya Sen

WITH SOME TWENTY BOOKS, INNUMERABLE LECTURES, AND SOME FORTY HONORARY DEGREES TO HIS CREDIT, AMARTYA SEN IS A THINKER NO ONE EXCITED BY IDEAS CAN IGNORE. HIS CAREER HAS SPANNED THREE CONTINENTS—ASIA, EUROPE, AND NORTH AMERICA (WHERE HE WAS THE FIRST NON-AMERICAN PRESIDENT OF THE AMERICAN ECONOMIC ASSOCIATION)—AND HIS WORK HAS COVERED SEVERAL OF THE CENTRAL ISSUES IN ECONOMICS. IN 1998 HE SURPRISED FEW ECONOMISTS WHEN HE WON A NOBEL PRIZE.

Sen is 'the conscience of the economics profession', says Nobel laureate Robert Solow, emeritus professor of economics at the Massachusetts Institute of Technology. 'Economists are very good at thinking about the "efficiency" of economic arrangements,' Solow explains. 'Since we are good at thinking about efficiency, we think about it all the time and write papers about minor differences in efficiency. Now efficiency has nothing to do with equity: an arrangement can be efficient while it starves some part of the population. Everyone is aware of the paradoxical character of this obsession with efficiency and inattention to equity. But most of us go on doing what we know how to do. Amartya has insisted that the profession should at least face squarely the ethical consequences of what it does. I don't think he pretends to have "solved" the problems of equity that arise. But he tries to get his economist readers not to walk by them without a glance.'

Sen is more than just an economist however. He has also published significantly in philosophy and regards his work in economics and philosophy as inextricably joined.

The urge not to specialize came from his childhood education. He was the son of a professor of chemistry and the grandson of a distinguished Sanskrit scholar (whose *Hinduism*, translated from Bengali by his grandson in 1960 and published by Penguin, remains a bestseller). His father was professor at the University of Dhaka, capital of what is now Bangladesh, where the young Amartya, born in 1933, spent most of his first eight years. Then he moved to live with his grandfather at the rural school and university founded by Rabindranath Tagore in West Bengal at Shantiniketan, about 100 miles from Calcutta. Although Tagore had died the previous year, his polymathic influence permeated everything. Like Dartington, the institution in England inspired by Tagore, Shantiniketan students were encouraged to feel that there was more to life than academic success. The emphasis was on the arts and on personal development; Sen enjoyed starting a night school to teach local Santhal villagers the three R's, thus foreshadowing his adult concern with literacy programmes. Shantiniketan, he says, 'was rather important in making me wide-eyed about having a variety in one's life.'

Soon after he arrived there, aged 8, he had an experience which gored him. 'A very skinny man appeared in our school compound behaving in a deranged way.' Over the days that followed, ten, then a thousand, then countless emaciated people tottered into the university campus. 'It is hard to forget the sight of thousands of shrivelled people—begging feebly, suffering atrociously, and dying quietly.' Meanwhile no one he knew personally, 'in the sense of belonging to the same class or family,' was short of food: 'they hadn't the slightest problem in continuing to live as they had before.' In Calcutta it was the same: he saw thousands of village people dying in the streets, while the life of the city was basically unaffected. This catastrophe, the great Bengal famine of 1943–4, with its shocking contrasts, was a strong spur to Sen's taking up economics.

All his adult life, Sen, a small, spare, energetic figure, has been restlessly on the move—both intellectually and physically (he is an indefatigable traveller). The beginnings of this inner drive can even be dated. In 1952, at the age of 18, while studying for his BA at the University of Calcutta, Sen discovered he had cancer of the mouth. He was treated with a massive dose of radiation, the maximum then permissible. 'It was an exercise in brinkmanship,' he says, which barred him from eating solid food for five months. He was told there was a high probability of a relapse within five years. 'Of course it affected my psychology. It was part of the reason for my always being in a rush.'

At Presidency College, then Calcutta's leading college, he received a superb technical training in economics and developed the talent for mathematical reasoning that has ensured professional respect for his work even from economists suspicious of his interest in ethical issues. Then, in 1953, he migrated to the University of Cambridge, where he gained his Ph.D. and became a fellow of Trinity College in 1957.

His subsequent institutional career has been as varied as his interests. He had a spell in India from 1963–71 at the Delhi School of Economics, and helped to make it India's leading centre for economics. Then he returned to Britain, to the London School of Economics, before moving to Oxford in 1977, latterly as Drummond professor of political economy and fellow of All Souls College. In 1987, he deserted the dreaming spires for the delights of Harvard Yard, where he was both Lamont university professor and professor of economics and philosophy—a combination which is perhaps possible only in America—only to return to Britain in 1998 as master of Trinity College, Cambridge: the first Indian, incidentally, to head an Oxford or Cambridge college.

Throughout, Sen has kept his Indian citizenship, unlike some of his economist colleagues. 'I love being in India,' he says, 'and in fact after my undergraduate days have never been away from India for more than six months at a stretch. But I also like the academic life in a more severe atmosphere, whether it be Oxford, the LSE or Harvard'. There is also the question of the specialized medical treatment he needs because of his radiation history, and the convenience of western living, and the fact that his wife, the historian of economic thought

Emma Rothschild, is also based in Cambridge and some of his children live abroad (he is divorced from his first wife, the Bengali writer Nabaneeta Dev, and his second, Italian wife, Eva Colorni, died in 1985). This bicultural arrangement allows him to immerse himself properly in the part of his work—two-thirds, he estimates—that does not directly concern India, while keeping in close contact with his Indian projects.

His work on social choice theory—how social decisions can take note of individual values—is part of that two-thirds. In the early 1950s, Sen read Kenneth Arrow's recently published *Social Choice and Individual Values* and became 'besotted' by it. Arrow investigated how, in democratic societies, different individual preferences can be combined into a collective judgement. His celebrated 'impossibility theorem' showed that only four apparently quite reasonable conditions (such as that there are no restrictions on the way in which people rank different alternatives) make the reaching of consistent collective judgements impossible. A well-known example is the paradox of majority voting whereby, in a majority vote for, say, the US capital, Washington D.C. would be beaten by Boston, Boston would be beaten by Chicago, but, in a vote between Washington and Chicago, Washington would win.

Sen both extended Arrow's impossibility theorem and, further, showed ways of resolving the conflict through the use of richer information about individual preferences.

In the standard economic view of preference, that of Paul Samuelson, 'the individual guinea pig, by his market behaviour, reveals his preference pattern'. In other words, if a person chooses a particular thing or action A, we can deduce that he gets more satisfaction from A. But this overlooks the possibility that the person chooses A because he believes it would be right to choose A, for one reason or another. Sen argues that economics needs to take more account of the complexity of motivation behind actual human choices. 'He stresses the ways in which group norms and the culture of a class or community can affect choices. For example, in choosing an insurance plan or making wage demands, people working in a particular job may be influenced as much by the attitudes of their fellow workers as by their personal preferences,' wrote A. B. Atkinson, warden of Nuffield College, Oxford, whose work on economic inequality has influenced Sen.

In various writings, especially *Collective Choice and Social Welfare*, published in the 1960s and 1970s, Sen introduced the ideas of rights and freedoms into social choice theory, and presented a way of formalizing this. It was 'pathbreaking' work, according to Arrow, but unlikely to excite the interest of a non-specialist readership. Gradually, however, Sen's thoughts focused on a problem of widespread interest: the cause of famines and the role of economic interaction in generating them. In 1981, he published a short monograph, *Poverty and Famine: An Essay on Entitlement and Deprivation*, and then, in 1989, *Hunger and Public Action* (written with Jean Drèze, a Belgian economist at the University of Delhi and close collaborator with Sen). Solow again: 'I think these two books literally changed the way intelligent people think about famine and widespread hunger. In many ways they may be Amartya's most important work. There is no highbrow economist's technique in them, but they ask the right questions and they are not put off by conventional answers.'

The issue, in a nutshell, is why famines occur even though there is no overall decline in food supply. In Bengal, for instance, in 1943, the supply of rice was 13 per cent *higher* than in 1941; but in 1943–4 as many as three million Bengalis starved to death, while in 1941 there was no famine. By careful analysis of data from the Bengal and other famines, Sen showed that the famine was not caused by an overall shortage of food, but by a shortage of practical capability among some people—mainly agricultural labourers—to obtain food. These people lacked income, or political weight, or social standing or other means to get sustenance. The word 'entitlement' in the book's title summed up the collection of powers which enables people to meet their needs.

One of the book's unusual features was Sen's understanding of social factors. In some families, for example, the male head of the household might claim food first, or boys would take precedence over girls when food was limited. The impact of gender relations on economic behaviour is a thread running through much of Sen's work. (His 1990 article, 'More than 100 million women are missing', in *The New York Review of Books*, has become one of his best-known writings.)

In the nature of academic life, Sen's case was not universally accepted. He cited evidence from Asian and African famines where, he said, food supplies had not significantly changed overall but where people starved because they lacked adequate entitlements. Some critics claimed, however, that food supplies *had* actually fallen in the cases Sen used; one even went so far as to accuse Sen of falsifying data. An acrimonious correspondence ensued. But today Sen's analysis is widely accepted, including by Oxfam; and the debate also had the effect of introducing Sen to Jean Drèze, whose critique of the 1981 book led to the jointly authored 1989 book.

Sen's comparison of famines in India, Africa, and China led him to what is undoubtedly an important conclusion. There have been no serious famines in India since 1943, during which period India became a democratic nation (from 1947). China, by contrast, suffered probably the worst famine in history between 1958 and 1961, when up to 30 million people perished. And Africa, in particular the Sahel, Ethiopia, and Somalia, has continued to suffer regular famine. The difference was public opinion, observed Sen: in India, it can exert pressure on the government through a free press and in other ways; in China and in many (though not all) parts of Africa, it cannot.

So the relief of poverty in a country, Sen maintains, entails much more than market-led policies designed to accelerate economic growth. It also, crucially, entails freedom. Referring to his most recent book, *India: Economic Development and Social Opportunity*, once more written with Drèze, Sen says: 'Jean and I have an indivisible view of freedom—that freedom in economic, political, and social matters is central for social change and progress, for economic progress and for development itself.'

This book, and Sen's work generally over the last decade, compares India with China, and Indian state (e.g. Kerala) with Indian state (e.g. Punjab). All have failed in one way or another to achieve the best policy combination. India has long enjoyed substantial political freedoms and in 1991 embarked on economic liberalization, freeing trade and investment after decades of bureaucratic controls. But continued neglect of social policy such as education and health undercuts these freedoms by making it hard for hundreds of millions of Indians fully to take advantage of them.

China, by contrast, has few formal political freedoms, but started to liberalize its economy in 1979 and has much higher literacy and life-expectancy rates than India as a result of effective government policy from 1949. Within India, the southern state of Kerala has achieved literacy and birth rates which compare very favourably with rich countries, but has a low per capita income. While a northern state such as Punjab is relatively backward socially—shockingly, half the girls aged 14–16 surveyed in north India had never been to school—but has had more economic expansion than Kerala.

What emerges from the accumulated work of Sen and his colleagues is more than an approach to development, or even a clever synthesis of different disciplines, mediated by economics. There are elements of a fresh political platform. Sen makes no bones about his leanings: 'I see myself as left of centre—decidedly so.' This fact drew predictable criticism of his Nobel award from one columnist in *The Wall Street Journal*. But he admits there are some 'unexamined prejudices' on the left. 'One is a basic mistrust of the market, the belief that somehow organizations which encourage people to operate through the market must be antisocial. There's no particular reason to assume that. After all, the original championing of the market came from radicals like Adam Smith—whose works inspired many of the intellectual leaders of the French Revolution. It's easy to see why. Insofar as respect for individual freedom is part of the radical tradition, the market is part of that because people should be able to exchange goods.'

'The other unexamined position is a kind of general belief that somehow fiscal prudence is an alien idea to the left. The government can spend money without being worried about how big the deficit is. That is a big mistake too. Inflation is not only a conservative worry. It can make the poorest of the poor go under'—as it did in wartime Bengal. 'So the idea that fiscal conservatism has to be conservative in the political sense is a mistake. Fiscal conservatism has to be part of a responsible left-wing position too.'

But before Sen begins to sound too much like New Labour, he adds, 'What I am also keen on, is not to lose sight of some of the things which make the left's position credible and important, namely, to recognize that the state has a role' and, further, 'to recognize the importance of social cooperation outside the market.'

Despite the commitment, the language is characteristically cautious. Amartya Sen is not given to dramatic statements and revolutionary theories of social change: he is too deeply aware of the diversity and subtlety of human societies. If that means that in certain aspects of his work he may be more of a gifted synthesizer than a theoretical innovator, this may be, like conscience, a strength as well as a weakness, depending on one's point of view. As Sen remarked in the conclusion of his 1991 Darwin lecture, 'On the Darwinian View of Progress', the publication of Darwin's great book changed the intellectual world irrevocably, without a doubt. But he went on to warn: 'A world view based on the Darwinian vision of progress can also be deeply limiting, because it concentrates on our characteristics rather than our lives, and focuses on adjusting ourselves rather than the world in which we live. These limitations are particularly telling in the contemporary world, given the prevalence of remediable deprivations, such as poverty, unemployment, destitution, famine, and epidemics, as well as environmental decay, threatened extinction of species, persistent brutality to animals, and the generally miserable living conditions of much of humanity. We do need Darwin, but only in moderation.'

by Andrew Robinson and Michael Prest

Things to Come

AMARTYA SEN

Breakthroughs are hard to predict. They are even harder to demand. But it is sensible enough to think of a shopping list of what we want. In fact, depending on our view of society, there is a connection between what we want and what we might end up getting. While Aristotle agreed with Agathon that even God could not change the past, he did think that the future was ours to make. In this sense, predictions cannot but link closely with what we intend to argue for, and ultimately, fight for.

In my shopping list I would certainly include more spread and consolidation of democracy. The century that is coming to an end has seen the establishment of the idea of democracy as the 'normal' form of legitimate governance. Normality is judged here not entirely by frequency, but rather the way a 'normal eye-sight' is defined.

It was common in the nineteenth century to ask whether such and such a country was 'ready' for democracy yet, or whether such and such a nation was 'fit' for democracy at all. There has been a sea-change in the way claims to democracy are judged now—at the end of the twentieth century. Rather than asking which nation is fit for democracy, the central question now tends to be how and in what way a nation can succeed in political, social, and economic life by means of democracy. The query is no longer about being fit *for* democracy, but about ways and means of becoming fit, as it were, *through* democracy.

There are reasons to feel somewhat optimistic about democracy. Democratic forms of governance have recently received much wider acceptance in the countries of Latin America and Africa, where—earlier on—scepticism about democratic politics had combined with the naked display of military power to overturn many democracies and to prevent the flowering of new ones. Also, the intellectual challenge to the acceptability of democracy that came from East Asia, with the diverting slogan that discipline was preferred over political freedom in 'Asian values' (allegedly traditional in Asia) has not won any intellectual victories, or managed to achieve much political acceptability of the kind that its advocates demanded. Even the major show-cases of non-democratic forms of governing such as South Korea and Taiwan have moved on to choose democratic forms of governance, with local support and evident success. The historical research on which the diagnosis of 'Asian values' was based was never particularly deep, but the shallowness of that piece of *ad hoc* intellectual history has become thoroughly clear in extensive works on past traditions in the Asian countries—in the east as well as in the south and the west of Asia.

What about Africa? The subverting of democracy in Africa, in the 1960s onwards, had much to do with the contingent fact of the Cold War, which was largely fought in territories far away from the Soviet Union and the United States. Any military strong man who was able to oust a democratic government in Africa could get instant support from the Soviet Union if it were anti-Western, and from the United States, if it were anti-Soviet. The dictators never lacked powerful backers in a world caught in an all-consuming Cold War. That war, happily, is now over and the 'world powers'—or whatever remains of that plural concept (now rather singular)—are no longer bending over to woo anti-democratic dictators to be lined up on their side in the battle against some global 'enemy'.

All this is a great relief. Furthermore, the intellectual climate of opinion is much more sympathetic to democracy today, both within the advanced economies of Europe and North America, and in the world of poorer countries in Latin America, Africa and Asia. There are good reasons for confidence that democracy may become not only the accepted form of acceptable governance, but also the accepted form of governing in practice.

The other item in my shopping list that I would like to identify is a fuller use of reasoning in social matters. This is, of course, the slogan under which a good deal of the intellectual battle of European

Enlightenment occurred. That was two centuries ago. The Enlightenment has been attacked quite severely in recent years. I believe
often quite unfairly. The claim that it was based on an arrogant belief
in the superiority of a very narrow mode of reasoning, and on an
intolerance of opposite points of view, is not easy to establish. That
diagnosis would correspond more closely to the immediate outcome of
the French Revolution (and the Reign of Terror) than to the intellectual under-pinning of the Enlightenment itself.

There is also an issue of cultural identity here. To think of the
European Enlightenment as a narrowly 'European' phenomenon
would be a mistake. Indeed it would be a *costly* mistake since this
diagnosis would make the Enlightenment look more sectarian than it
ever was. The leaders of the intellectual traditions of European
Enlightenment not only drew on the Renaissance (a quintessentially
European and largely Italian phenomenon), but also absorbed the
influence of Chinese science and engineering, Indian and Arabic
mathematics, and other contributions from across the world (which by
then had become 'native' in Europe itself). The European Enlightenment was staking a claim on behalf of the intellect of humanity, not as
a partisan of a narrow 'European point of view'.

It is sometimes claimed that 'enlightenment' is a distinctly European notion. This, of course, is just rubbish. The concept of
enlightenment, in a general form, has been invoked in early writings in
Sanskrit, Pali, Chinese, Arabic, and other classical languages with a
great deal of insistence and reach. To illustrate, it may be worth recalling that Gautama Buddha was given the name 'Buddha' meaning 'the
enlightened one', and this reflects the central claim to excellence that
his followers attributed to him. No matter whether his studied
agnosticism appeals today to European theists or Oriental funda-
mentalists or not, or whether his insistence on reasoning about every
issue rather than blindly following established tradition is seen to be
acceptable in the contemporary West or in radically conservative non-
Western societies today, these are the claims that were put forward

most forcefully by Gautama Buddha two and a half millennia ago. Phenomena of this kind have occurred again and again in the history of different parts of the world. To suppress reason in favour of faith and unreasoned belief amounts not merely to questioning our own ability to think, but also to denying the analytical heritage of human civilisations, of which we are the inheritors.

My shopping list, as may be clear, is not unambitious. But nor is it particularly over-demanding. Indeed good use of reasoning and democracy can help to consolidate both reasoning and democracy. The ends and the means are not in conflict with each other.

Further Reading

Sen, Amartya, *Collective Choice and Social Welfare* (Amsterdam, 1970).
—— *Poverty and Famines* (Oxford, 1981).
—— *Inequality Reexamined* (Oxford, 1992).
—— 'Rationality and Social Choice', presidential address, American Economic Association, *American Economic Review*, Mar. 1995.
—— *Development as Freedom* (New York, 1999).
—— 'The Possibility of Social Choice', Nobel lecture 1998, *American Economic Review*, July 1999.

Elaine Showalter

WHEN HER PARENTS DISOWNED HER—SENDING HER SISTER AROUND WITH A LIST TO PACK UP EVERYTHING THEY HAD GIVEN HER—FOR MARRYING ENGLISH, THE NON-JEWISH ACADEMIC WHO REMAINS HER HUSBAND TODAY, ELAINE SHOWALTER FOUND IT 'LIBERATING'.

'My parents would have given me a lot of flak about doing a Ph.D.,' she says. 'They would have given me flak about having children and working. There were many things I did ... this was the 1960s—they wouldn't have liked the way I dressed. It was just much easier not to have to deal with them.' The break with her parents also helped Showalter develop her life alongside that of the women's movement, which she joined virtually as soon as it started.

Now, at 58, Avalon professor of the humanities at Princeton University in New Jersey and a celebrated feminist literary critic and historian of science, she has been determined to make her own way, to earn her own money, and to break free from the conventions which dogged her early years.

The first person in her family to attend university, and told she was 'too bright for a Jewish girl', she studied first at the women's college Bryn Mawr in Philadelphia, before graduate work at Brandeis, and a Ph.D. on Victorian women writers at the University of California, Davis. Bryn Mawr was, she says, 'the most oppressive, conventional, suffocating, genteel, snobbish, dire sort of place'.

When she later started teaching at Douglass College, a woman's college at Rutgers University, she was determined to change things for her students, fighting to give them better employment opportunities as well as practical help such as counselling and childcare. 'I was very fired up to try to make up to them for what I had lacked,' she says, 'all the things I had missed.'

Through her academic work she was already striving to make up to neglected women writers some of the attention they had been deprived of over the years.

In *A Literature of Their Own: British Women Writers from Brontë to Lessing*, published in 1977, she argues that by examining a wider range of women writers 'the lost continent of the female tradition has risen like Atlantis from the sea of English literature'. There *is* a female tradition of literature, she argues but it had been subsumed by dominant male definitions of what was culturally important. Showalter raised the profile of writers such as Mary Braddon, Sarah Grand, Rose Macaulay, and Rebecca West whose works were out-of-print at the time. In the process, says Isobel Armstrong, professor of English at Birkbeck College and editor of several anthologies of women's poetry, 'she altered a mindset'.

As her work became better known, she was invited to argue her case at conferences and meetings across America, alongside some of the most eminent male literary theorists of the time. Armstrong pays tribute to her courage in speaking out in the face of tough opposition, forcing the literary establishment to take women's writing seriously.

This was when Showalter coined the term 'gynocriticism' which 'begins at the point when we free ourselves from the linear absolutes of male literary history, stop trying to fit women between the lines of the male tradition, and focus instead on the newly visible world of female culture.' It was a calculated move to match the language then being invented by structuralists and deconstructionists and it helped lend her own theories the same kind of *gravitas*.

She has nevertheless been criticized since by feminists for not being sufficiently theoretical. Certainly, her style of writing, unambiguous and lucid, was very different from the difficult prose of some Continental theorists. The feminist critic Toril Moi attacked her for being too materialist and pragmatic, for failing to take account of 'the text as a signifying process', as something with symbolic weight as well as practical meaning, for being insufficiently postmodern. But Showalter says she has also been criticized for being too theoretical and dismisses attacks from both sides as 'part of the territory'.

As she was helping change attitudes to women both in life and literature, past assumptions still dominated Showalter's own life.

After her marriage, aged 22, she left graduate school a year into her course at Brandeis to follow her husband, found herself turned down by colleges with a no-women hiring policy and ended up teaching at Douglass, a women's college, with a baby. 'At the beginning, I really didn't have any expectations or ambitions and I made certain decisions lightly—such as to have kids—because I had no idea I could have any career at all,' she says. She had her son and daughter and became an academic because she could not see many other choices open to her. Anyway, she enjoyed being a mother and liked teaching. 'I thought I would be able to teach at the lowest level. I never had any idea I would be successful at it.'

By this time, she had made her first trip to London and had instantly fallen in love. 'I had the *coup de foudre*,' she says. 'I really felt England was the place I was meant to be.' She lived in London for two years during the 1970s and still spends part of every year there, usually concentrating on the more scientific side of her research interests.

Her early trips to London coincided with research for her book *The Female Malady: Women, Madness and English Culture 1830–1980* and she discovered the Wellcome Institute, then, she says, in its 'golden age'. She rubbed shoulders with medical historians Roy Porter, Bill Bynum, and Chris Lawrence, breaking away from women's literature and gynocriticism into the history of science, medicine, and madness.

Again, the spark was personal experience. 'I came of age in the 1960s when it was very common to see women having all kinds of breakdowns,' she says. Her mother, who died in autumn 1998, suffered a series of hysterical illnesses—'very dramatic'—while a girl she was close to at college had a severe breakdown and mutilated her face.

This had a profound effect on Showalter, who was told by the girl's psychiatrist that when a woman mutilated her face he considered it worse than suicide.

Showalter herself had her own psychological problems as a teenager—'relatively primitive', she says. 'I'm a pretty sturdy specimen.' But the difficulties she had were exacerbated by 'the total stupidity of the psychiatric establishment and the very conventional, traditional society.' Her parents, at one point, made her see a psychiatrist who then filed a report on her to them. 'It was really about surveillance,' she says. 'It was not about help. It was not about understanding.'

The Female Malady is a sane, but passionately argued, look at the treatment of female madness over the last century, suggesting that 'we can expect no progress when a male-dominated profession determines the concepts of normality and deviance that women perforce must accept'.

She updated the idea in *Hystories*, published in 1997, which argues that many modern-day maladies and fears are cultural symptoms of stress, suggesting too that epidemics of hysteria seem to peak at the end of centuries when people are particularly alarmed about social change. 'Patients learn about diseases from the media, unconsciously develop the symptoms, and then attract media attention in an endless cycle,' she says.

The book, which equated chronic fatigue syndrome and Gulf War syndrome with fears of alien abduction, caused a storm. She received regular hate mail and was accused of compliance with right-wing conspiracies. Far from being scared, she claims she regarded the attacks as good free publicity. 'It was a challenging book and I wrote it to be challenging,' she says. 'Also, I was writing at a stage in my life when I could feel very brave about it. That is partly because I was 50 when I wrote it.' By then, she says, she was 'eager for controversy'.

Also, she has not a moment's doubt that she was right. Her method of working is meticulously to build up mountains of evidence before settling down to write. She knows a lot about Gulf War syndrome, a lot about chronic fatigue syndrome and is convinced that they will finally be proved to have psychological rather than physical causes.

But her motive in writing the book was to highlight hysteria generally and she tried to confront everyone with their own prejudices. For example, when she appeared on radio phone-in programmes in the United States with large African-American audiences she would find plenty of support for her views on chronic fatigue syndrome, which almost exclusively affects white people. But then she would ask callers whether they believed the CIA were selling drugs in California or whether Kentucky Fried Chicken made black men impotent. Such questions drew the callers up short. 'I have become a much more rational, scientific, sceptical person than I was before I wrote the book,' she says. In a way, it was the book she needed least courage to write, she says, because it had very little to do with her.

For her next book, she will draw more than ever on her own personal experience, exploring the lives of feminist intellectuals from the last century to the present day. Some of them she has met. Many have had an impact on her own life and work.

Among them, is Princess Diana—not, Showalter concedes, an intellectual but someone who has reflected on womens' lives and had a deep psychological impact on the women of her time. 'I'm interested more in women who at some point developed a model of what it would mean to live the fullest possible life as a woman,' she says.

Her own life, free from its early restrictions, is also becoming fuller. One of the great things about being a woman now, she says, is that life does not end at an arbitrary cut-off point. In the second part of her life, as well as being a grandmother, she has been able to fulfil an early ambition to be a journalist. In 1996, she wrote a TV column for the mass market American tabloid paper *People*. She has also written book reviews and cultural criticism for a newspaper readership.

For her, 'The women's movement succeeded beyond our wildest

dreams.' There is still more to be achieved, she says. Childcare, for example, could be improved and care for elderly relatives still falls mainly on women, who need more help.

Showalter has direct experience of this herself. While her sister looked after their mother for the twenty years before her death, she was responsible for meeting the costs. 'Yes,' says Showalter. 'I made it up with my mother just in time to pay the medical bills'—a liberation of sorts.

by Harriet Swain

Optimism v. Pessimism

ELAINE SHOWALTER

In the 1890s, both feminists and Americans were optimistic about the future, and expressed their hopes and desires in utopian stories and novels. Edward Bellamy's *Looking Backward*, a vision of the year 2000, sold over a million copies. While Europeans pessimistically embodied the apocalyptic spirit of the *fin de siècle* in dark figures like Dracula, Mr Kurtz, or Dr Moreau, Americans in 1900 were reading about a benevolent humbug called the Wizard of Oz, who used science and technology to create the illusion of happiness. American feminists were the boldest of all, imagining female social utopias where women had not only legal rights but also complete freedom of sexual experience and initiative.

As an American feminist I tend to be an optimist too, but women's fantasies and hopes for the future depend on scientific and technological progress, as well as political enlightenment. It now looks as if the battle for legalized abortion has been won, despite the continuing resistance of religious right-wingers and anti-abortion terrorists in the USA. Progress is underway in the detection, diagnosis, and treatment of breast cancer, and I expect ovarian cancer to be cured in the twenty-first century. Attitudes towards pregnancy and childbearing continue to develop and improve, especially in the area of infertility, problem pregnancies, and maternal care. I hope that the dramatic trend towards the involvement of fathers in parenting will continue.

But in many areas social services have not caught up with social change, and in the next century, feminists will demand that governments pay attention to the practical needs of working families by providing high-quality childcare, early education, and more flexible working shifts. Women's lives and opportunities are also curtailed by responsibility for the elderly, and for every John Bayley tenderly nursing a beloved wife with Alzheimer's, there are a dozen women caretakers of the aged and infirm. I hope that research on Alzheimer's will relieve us all of this scourge, and I would like to see much more realistic social and legal attitudes towards dying.

I am more pessimistic about psychiatric and psychogenic disorders. The twenty-first century will also bring new paranoias, new hysterias, new conspiracy theories, and new imaginary illnesses; and if history is a guide, women will be the majority of believers and sufferers. With its mythology of the saved and the damned, the millennium fuels a volatile mix of psychological vulnerability and uncertainty. Women have always channelled their psychic distress into conversion symptoms, have flocked to faith healers, quacks, and alternative therapies. I would like to see reason win out, and the breakthroughs in the treatment of depression help us recover the energy, both feminine and masculine, wasted in melancholy. But there is a side of me that agrees with Woody Allen: 'I have seen the future, and it is very much like the present, only longer.'

Further Reading

Showalter, E., *Sexual Anarchy* (London and New York, 1990).

—— *Hystories: Hysterical Epidemics and Modern Culture* (London, 1997).

—— *A Literature of Their Own*, revised edn. (Princeton, 1999).

Peter Singer

I T IS ONE OF THE GREAT BRITISH ECCENTRICITIES; OUR MOIST-EYED LOVE OF ANIMALS. MORE AND MORE OF US COUNT OUR DOG OR CAT AS OUR PRIMARY COMPANION, WELCOME THE SENTIMENTALITY OUR PETS INDUCE IN US, ARE PROUD TO CONSIDER OURSELVES ANIMAL LOVERS.

But, in the eyes of the man who has done more than most to promote the rights of animals this century, the British people are far from being a true nation of animal lovers. According to philosopher and writer Peter Singer, those who dote on a sleek Siamese, or selflessly rescue a dog from the pound, have more in common with slave owners than they would care to admit. The Australian animal rights campaigner doesn't even like the term pet. 'It's a put down,' says Professor Singer dismissively.

There is nothing misty-eyed about Singer's commitment to animal rights. 'I wasn't ever particularly emotional about animals,' he asserts. 'I was never an animal lover', didn't have any kind of 'companion animal'. He stresses this last phrase, with an inflection that makes it sound even more patronizing than the politically incorrect 'pet'. So how was Singer, now professor of bioethics at Princeton University, convinced that this was a campaign that needed his support? Not by emotion, but by rational debate, he says.

'I was persuaded by rational arguments that what we were doing to animals was an injustice, a wrong, an exploitation. Later, when I read vivid accounts of exactly what happens when shampoo is squeezed into a rabbit's eyes or whatever . . . then yes, I got emotional about it, then.'

He believes that emotion may be a powerful impulse, but that it is no basis for ethical decisions. His book *Animal Liberation*, written when he was 29 and now the bible of the animal rights movement, is a closely reasoned justification of the rights of animals. 'Nowhere in this book,' he emphasizes, 'do I appeal to readers' emotions where it cannot be supported by reason.'

He prides himself on the rigour of his thinking: 'I like to think that I have put pressure on people working in the field of bioethics to make their arguments more rigorous, not just to mouth platitudes about "the intrinsic dignity of human beings" or "the inherent worth of human life" but to actually show what they mean, why they think that humans inherently have dignity and animals do not.'

In the thirty or so years Singer has been at the forefront of animal welfare issues he has seen enormous progress. 'In the 1960s there was no discussion of a whole lot of aspects of animal suffering that are now very well known. Until Ruth Harrison's book *Animal Machines* was published in the 1960s no-one knew what animals were suffering in factory farms. I didn't. In the late sixties, when I was a student at the University of Oxford, I had never heard about battery cages. I never thought about the fact that cosmetics were tested on animals. Bottles on sale at chemists' shops were never labelled "Not tested on animals".'

In recent years, Singer has championed the cause of great apes with tangible success. In *The Great Ape Project*, co-edited with Paola Cavalieri, he proposes that great apes, as intelligent beings with rich and varied social and emotional lives, should have the same rights and protection as people. 'If we are looking at extending rights to animals, then the case for the great apes is even more overwhelming than it is for other animals, because the great apes have so many of the capacities that we believe entitle us to have rights,' explains Singer. 'They lead rich emotional lives, they form long-standing relationships, they show signs of self-awareness. They show grief, they play; when taught sign language they tell lies; they plan for the future, they form political coalitions; they reciprocate favours.'

He goes on: 'If you want to argue that all humans have certain basic rights, including infants or the intellectually disabled, then the case that great apes ought to have the same rights is clear: great apes are the intellectual equivalent of 2-year-old children or some people with intellectual disabilities.'

Already, in the United States, the National Institutes of Health has issued a paper stating that scientists should not kill great apes that are no longer needed as subjects in medical or scientific experiments. (Though Singer is swift to point out that the US government's recommended cage size for a single adult chimpanzee is still 5 ft × 5 ft × 7 ft.) In Britain, the government has announced that it will no longer give permission for great apes to be used in medical research.

And in New Zealand supporters of the Great Ape Project are pressing the parliament to amend the country's animal welfare act to give great apes three basic rights: the right to life, the right to liberty, and the right to freedom from torture.

Singer hopes that eventually these rights will be incorporated into a declaration by the United Nations. 'All I want is that we protect apes in the same way that we protect vulnerable humans—that we show concern for them, reject the idea that they can be owned and, where they are at risk, we appoint guardians to safeguard their interests.'

Yet what is exciting the media on the occasion we meet are not Singer's views on animals, but remarks he made several years ago about the right to life of handicapped babies. Hardly had he stepped off the plane for a week of lectures in London than newspapers and television programmes were presenting him as a baby killer, apparently more concerned about animal welfare than the rights of the most vulnerable of his own species.

In fact Singer's position on handicapped babies follows directly from the principles that underlie his views on animals. Currently, babies born severely disabled often die a lingering death because in most countries euthanasia is illegal. The infants' food, drugs, and life-support are withdrawn, and they die slowly and in a manner that can be extremely distressing for them, their parents, and the nursing staff.

Singer argues that doctors would be justified in administering a lethal injection (with parental consent) because a swift, painless death is in the child's best interests.

In Singer's world-view we do not have an inviolable right to life just because we are human. Not all lives are of equal worth—because degrees of self-awareness, intelligence, and the capacity for meaningful relationships with others make some lives more valuable than others. The presence or absence of capacities such as self-awareness are relevant to decisions about whether or not to end life. A chimp, a dog, even a pig has more self-awareness and capacity for meaningful relations than do some severely intellectually disabled children.

It is our status as sentient beings, able to feel pain, to suffer, and to anticipate death, that matters most to Singer. He argues that it is time to abandon the Western, Christian-based ethic of the sanctity of human life. The Old Testament Commandment 'Thou shalt not kill' says nothing about not killing animals. Singer considers this 'speciesist'—an example of discrimination against animals based on the belief that humans are more important—and thinks it is as wrong as racism or sexism.

More practically, Singer points out that medical technology has advanced to the point where doctors, lawyers, and parents already make life-and-death decisions. This reality conflicts with the traditional Christian ethic to preserve life at all costs: the result is all too often 'tragic farce', Singer says.

His 1994 book, *Rethinking Life and Death*, contains a discussion of a legal ruling the previous year that the Liverpool football fan Anthony Bland, a victim of the 1989 disaster at the Hillsborough football stadium, could be disconnected from his life-support machine. In the book, Singer also described the case of Rudy Linares, a 23-year-old Chicago house painter who held nurses at bay with a gun while he shut down the respirator keeping his 8-month-old son alive. Linares cradled his son until the child died half an hour later. Then, crying, he gave himself up. 'He acted against the law and the sanctity-of-life ethic, but his impulse was in accordance with an emerging ethical attitude that is more defensible than the old one and will replace it,' Singer wrote.

As a utilitarian philosophy Singer's argument may be unassailable, yet many are unable to accept the logic of his conclusions. Dr John Wyatt, professor of neonatal paediatrics at University College London, says that there is all the difference in the world between letting people die and killing them. Singer replies: 'To me "letting nature take its course" does not make a lot of sense in a neonatal intensive care unit where one is making deliberate decisions to allow a child to go. If you work in high-tech medicine, you have to take responsibility for making life-and-death decisions. Dr Wyatt is not finding a way to prolong life to the last possible moment, he's making a decision at some point to pull back (from treatment). The crucial question is when that decision is justified rather than how exactly death comes about.'

In *Rethinking Life and Death*, Singer offers an ethic that 'looks at the quality of life and does not draw sharp lines between humans and animals as such'. He suggests that if we are to find fulfilment in life, developing a coherent ethical framework for the modern world is imperative. Is there anything still to live for apart from self, money, love, and family, he asks in *How Are We To Live?*, a book published in 1993. 'The possibility of living an ethical life provides us with a way out of this impasse. To live ethically is to reflect in a particular way on how you live and to try to act in accordance with the conclusions of that reflection.'

This pragmatism has been a hallmark of Singer's career. When he decided to abandon the law part of his law and philosophy degree at Melbourne University, he knew he would never be happy in an ivory tower. 'I have always felt that philosophy would not be worth doing if it did not have some impact on the rest of the world. In philosophy classes when people talked about how we know there's a desk here, I thought "that's a really interesting intellectual puzzle, but I'm not going to spend the rest of my life thinking about it".'

Singer is optimistic that, in the long run, reason will prevail over emotion and instinct in ethical decision-making. He has seen his approach bear fruit in animal liberation: 'People still come to me and say "your book changed how I live". That's great because it shows the power of rational argument.' In one hundred years, he thinks, we will have resolved the issue of voluntary euthanasia, will have given great apes a new moral standing, and we will be using genetic advances in many beneficial ways.

His mind goes off on a different tack and he starts to speculate on where globalization may lead. It may mean the end to national controls over genetic engineering or more economic equality between individuals. But then he stops himself. 'I'm talking off the top of my head.' He has to think his position through. If the rest of us did the same—would we agree with him about killing babies?

by Kate Worsley

A New Ethics

PETER SINGER

Since my field is applied ethics, and I think that many of our conventional moral judgements are mistaken, there are quite a few changes that I would like to see take place in the twenty-first century. I hope for breakthroughs in three major areas:

First, our attitudes to non-human animals. In the last quarter of the twentieth century there has been a marked change in the extent to which the suffering of animals has been taken seriously. This change is particularly evident in Europe, where it has led to a sharp rise in the number of vegetarians and vegans, and to political moves like the European Union's decision to phase out the standard battery cage for laying hens by 2012.

This is all positive, but I hope it will gather momentum so that by the end of the twenty-first century the interests of non-human animals in living a good life, free from pain and suffering, will everywhere receive the same consideration as that which we give to similar interests of humans.

Animals will no longer be mere property, but legally recognized as sentient beings, with rights that can be enforced in the courts by guardians acting on their behalf. The intensive confinement of animals indoors will have been ended, and hopefully, at least in developed countries, people will not use animals for food at all. Then people will look back on the way we treat animals now as we look back on the era of the slave trade.

Second, our attitudes to our fellow-humans in less fortunate parts of the world need to change. While people anywhere in the world are dying of hunger or easily treatable diseases, I'd like to see us take a much more negative view of others living in a luxurious, profligate way, while not contributing significantly to helping the poor. It is even worse when luxurious lifestyles cause greenhouse gas emissions that change the climate of our planet, with disastrous consequences for those least able to withstand them. We need to take responsibility for all the effects of our actions—and for the effects of our indifference as well.

Third, I would like to see changes in the way we handle life and death decisions in medicine. At present we are in a halfway house, still ruled by the vestiges of a Christian ethic that most of us do not really accept, but unable to take the step to something more coherent. This would mean the legalization of voluntary euthanasia and physician-assisted suicide, and a more honest and open attitude to life-and-death decisions in the case of patients in a persistent vegetative state, or newborn infants with severe disabilities.

The common thread in these three changes can be summed up very briefly: reducing avoidable pain and suffering. Since, even with the best of luck, I will not live beyond the first third of the twenty-first century, I do not expect to see all three breakthroughs. The first two are very far-reaching, and if they are achieved within the century, that will be remarkable enough. But the third is already upon us, and could come quite soon after the year 2000.

Further Reading

Singer, Peter, *Practical Ethics*, 2nd edn. (Cambridge, 1993).
—— *Animal Liberation*, 2nd edn. (London, 1995).
—— *Rethinking Life and Death: The Collapse of our Traditional Ethics* (Oxford, 1995).
—— *How are We to Live? Ethics in an Age of Self-Interest* (Oxford, 1997).

Dale Spender

'**W**E'RE ON THE VERGE OF A HUGE EDUCATIONAL REVOLUTION,' OBSERVES DALE SPENDER GLEEFULLY. THE DIGITAL AGE OF COMPUTERS AND THE INTERNET IS REPLACING THE PRINT AGE OF BOOKS IN WAYS WHICH WILL AFFECT THE WHOLE PRACTICE OF LEARNING AND STUDY. THE BRITISH GOVERNMENT'S CURRENT EMPHASIS ON TESTING LITERACY IN SCHOOLCHILDREN IS, SHE THINKS, 'MY GENERATION'S RESPONSE TO LOSING CONTROL'. SHE GOES ONE STEP FURTHER. 'LITERACY? IT'S PRINT LITERACY. YOU COULD ACTUALLY ARGUE THAT IT'S A DEVELOPMENTAL DISADVANTAGE.'

The Australian academic and writer Dale Spender, 55, is, by her own admission, a rebel and courts controversy. When I ask her why Australia has produced such influential feminists and whether it is because the men are so macho, boasting sheilas and surfing big rollers and braving sharks, she corrects me. 'Oh, it's because the women are so loud. We're not afraid to shout and so whatever we say really gets noticed.' She is certainly one of the loudest and most noticeable of them all. For our lunch at a very quiet and very English restaurant in the heart of Covent Garden, she is dressed from head to toe in purple, even including shoelaces. 'Everything seems so small,' she confesses as we first sit down. 'I feel I am going to knock something over.'

Her tremendous warmth and enthusiasm are infectious. Although we are supposed to be talking about the Pandora reissue of her seminal book *Man Made Language*, first published in 1980, we end up spending much of the lunch discussing the implications of the Internet, which is her latest source of excitement. She has just come back from the first conference held by Nottingham-based trAce, the Online Writing Community, where she gave a speech on 'Digital Arts: Breaking the Boundaries through Online Authorship'. Writers are worried by the Internet because it is interactive and they therefore have limited control over how they are read and what happens to their texts. Spender told them how they could use the web to their advantage: 'The printing press was around a long time before the novel emerged. Maybe there are new creative genres that are interactive. I think it is a huge challenge creatively.'

The wider cultural consequences of the Internet have not yet been acknowledged, Spender feels, and she is preaching them with messianic zeal. The World Wide Web changes the whole nature of communication, which has always been central to Spender's interests. It means looking rather than reading, it means making connections (or playing) rather than studying a finished text, it means being able to access information at any time you want. 'Print literacy is about following an argument. It's about middle-class values—postpone the gratification until you get to the end. There's no ending online. There's no closure, no linear basis. It's about bringing it in, checking it out, constantly evaluating.'

To help teachers cope with the great technological changes and the very different demands of digital literacy, Spender has developed an online professional development course for educators. She has also persuaded the University of Queensland, where she is a professor, to replace their library with a 'cybrary,' complete with five hundred computers and beanbags instead of desks. Indeed so involved with the possibilities of the Internet has she become that she regularly advises the Australian government about the latest innovations in the 'Information Superhighway' and tours the world giving keynote lectures on the wonders of the World Wide Web.

What amazes and distresses Spender particularly is that the digital revolution is passing universities by. While the rest of the world is 'nattering on the net' (the title of her latest book), universities are being marginalized. But surely, I argue, academics use e-mail more than anyone else? 'They are not making use of the media in the way you can use online,' she says. 'They are just putting ten-year old lecture notes on a website, which I just say is making the students do the photocopying.'

Indeed, she sees universities as part of an old order which is resisting change hard but which will eventually be overtaken or left outside the information revolution. 'In the past there was a stable body of knowledge which universities jealously guarded as custodians or gatekeepers and that's not how information works any more.' In future, Australian students will access information, generated perhaps in America, online at home, after work, as a social activity. Universities will no longer be centres of information, she thinks, because people will not need books or physical teachers. 'Students will vote with their virtual feet.'

All the cyberspace wizardry might come as something of a surprise for someone expecting the classic radical feminism which made Spender's name. But actually her interest in the possibilities of the Internet is not so far removed from her concerns in *Man Made Language*, since it also focuses upon the impact of the social environment upon communication. In fact, in books ranging from *Man Made Language* through *Learning to Lose: Sexism and Education* to *Men's Studies Modified*, she has always looked at the ways in which institutions are biased in favour of certain types of knowledge and communication and collude in silencing others. As she says, 'it's the changes in the nature of information that interest me. It's always been that. Even with feminism, it was: if you added women to the cultural traditions of information and knowlege, how much of what was already established became primitive and wouldn't work.' Yesterday's sexists, she maintains, are today's printists.

Man Made Language was the product of Spender's years in feminist groups in London in the 1970s. She had married and divorced in Australia and came over to London to reinvent herself and to study for a doctorate in socio-linguistics at the University of London. It was a hugely exciting time when women threw out their old clothes and their old ideas and femininity was rigorously examined and evaluated and mostly discarded. 'There was tension almost as soon as I got here', she remembers, 'because I said with almost absurd naivety that all my supervisor's research had been based on boys and so all the stuff about working-class language patterns had no women in it.' Determined to prove that gender played a significant factor in daily conversation, she systematically taped staff meetings, seminars, social gatherings, and meetings in the student bar and analysed the recordings.

The book that grew out of this research maintained that men and women talked differently. The way that men used language was considered the norm and prescriptive; the way that women used language was considered deviant and has consistently been denigrated over time, with terms such as 'gossip' or 'natter'. Men controlled language and, in Spender's words, 'it worked in their favour'. The common myth was that women talked all the time but in fact, on Spender's empirical data, it was men who dominated any conversation. She argued that linguistics as a discipline had denied the experience and practice of women, and men's domination of language was directly connected to their wider social and political oppression of women. 'Both language and material resources have been used by the dominant group to structure women's oppression and they are interconnected,' Spender wrote. 'One cannot be transformed without the other if women are to be liberated and patriarchy is to be prevented from persisting.'

The solution, according to Spender, was to study women's alternative way of speaking and to form consciousness-raising groups (known as CRs) and single-sex and mixed-sex conversation groups so that women's particular language practices could be validated.

Man Made Language was, according to literary critic Professor Janet Todd, exciting and important: 'What she says about the way language shapes one's consciousness and the sense that one can't easily get away from it is still current today.' But some people, particularly from the world of linguistics, criticized her for not knowing enough about linguistics and not using objective sample data. It was said, for example, that she did not have 'a control group' for comparison. 'There are pedantic people around who resent her bubbliness,' says Janet Todd.

Spender's interest in the difference of women's language led her to become concerned with women's writing as well as their speaking. Like Elaine Showalter, professor at Princeton University and many of the pioneers at Virago Press, she became involved in the recovery and publication of lost and forgotten texts by women. A series of important anthologies edited by her came out in the late 1980s: *Mothers of the Novel: One Hundred Good Women Writers Before Jane Austen* (1986), *The Penguin Anthology of Australian Women's Writing* (1988) and, co-edited with Janet Todd, *An Anthology of British Women Writers from the Middle Ages to the Present Day* (1990). The anthologies provided the raw material for the emerging subject of feminist literary criticism, but—since Spender is always keen to show how language and literature are crucially connected to the wider social reality—she also believes that they struck a blow for women's liberation. 'To reclaim and revalue women writers men have removed is to do more than challenge a biased version of literary history,' she argues. 'It is to take a political stand and to challenge the propaganda of a dictatorship.'

Twenty years on does Spender think that the arguments in *Man Made Language* are still current? 'I think there are areas where men are still treated more seriously as speakers than women,' she says. 'Basically men think that a successful conversation is one where they have the floor and women think that a successful conversation is one where everyone has a turn.' Indeed she thinks that one of the interesting purposes of the reissue of *Man Made Language* is to explore which practices have changed since 1980 and which have remained the same. 'In a sense if they are still happening, they are more shocking,' she says.

But she does feel that things have moved on from the dogmatism of late 1970s feminism. *Man Made Language* is unabashed in its use of terms like 'sexism' and 'patriarchy' and 'women's oppression' and very clear-cut in its polarization of men's language and women's language. While Spender is unhappy about the political implications of the change of terminology from 'feminism' to 'gender studies' ('gender studies doesn't have the same notion of power as women's studies—like "get it back"'), she admits and welcomes the fact that we now recognize multiple realities and greater variety. Next time, she would acknowledge that she was only writing about white women's language and would expect a black woman to write about their very different experience. And she would not want to replace a masculinist view of the world unthinkingly with a feminist one. 'I hope that all of us have moved to a much more postmodernist context, which is about the questioning of authority and how truth is made and where does that information come from,' she says.

If there is a power struggle going on now, it is not so much between the sexes as between the ages, she argues. It is young people who know how to work the technology and surf the net, while it is her generation—the babyboomers—who are attempting to maintain the supremacy of print culture. 'It's the first time in history that the younger generation have known more about the communication medium than the older generation and the older generation is being punitive in making the kids jump through the hoops they had to jump through.'

Spender's wide intellectual curiosity and forthright opinions are exhilarating. By the end of lunch we have covered a bamboozling array of topics. We have discussed Bill Clinton (he's offensive but witch-hunted), hairstyles (henna was OK in the 1960s but not bleach), divorce (men are disappointing), Samuel Johnson's bluestockings (silenced by literary history), whether she burnt her bra (she didn't), and Rebecca West (she interviewed her when West was in her eighties). Spender has won over the restaurant as friends, although she seems almost disappointed that she hasn't knocked anything over, at least literally if not metaphorically.

So is she looking forward to the revolution she predicts or would she ever want to abandon the iconoclasm and police the Internet? 'Our education system and the print system are about control and I don't think you can have that any more. Maybe there'll be days when I think that's terrible and maybe there'll be days when I think it's a good thing but I certainly think there's going to be a lot more anarchy and a lot less order, and that's the medium as well as the culture.'

by Jennifer Wallace

A Basic Human Right

DALE SPENDER

Throughout the Middle Ages, reading was regarded as such a specialized skill that only a talented few were considered capable of becoming readers. But then along came print—and extraordinary information opportunities. And it was not long before society was divided into the information haves (those who could read) and the information have-nots.

Those who were interested in social justice recognized that being able to read and write was such a source of empowerment that they wanted everybody to be literate. So in the nineteenth century there were great campaigns to promote mass literacy, and to make reading and writing for all a legal requirement. As the price of books was beyond the means of many, a system of public libraries was set up so that all could have access to print. This was a literacy revolution, often referred to as 'the democratization of reading'; it certainly promoted the democratization of society. Literacy has since been decreed a human right.

Now we face much the same challenge again. Just as the information medium once changed from manuscript to print, now it is changing from print to digital. And just as those who were illiterate in a print based society have been kept 'out of the loop' and unable to participate fully in the community, so now are computer illiterate people facing a similar fate as 'outsiders'.

It's not sufficient for developed societies to ensure that some of our most talented people become 'knowledge workers'. It's not acceptable to create an ever widening gap between the information haves and have-nots. The challenge for us now is to ensure that all members of society are not only computer competent but that they have their own computer connections as they once had books.

During the industrial age governments of every shape and size have spent incalculable amounts on the infrastructure. Roads, rail, sea ports, and airports have been built (again and again) to get the goods to market and the workers to work. We have understood the need for such enormous expenditure; the prosperity of the nation has depended on it. But we are agonizingly slow to recognize that in an information society, the information infrastructure is as basic and necessary as roads and rail have been in the last century. Only with a computer—and a connection—can the workers get to work and deliver the services that are part of the information economy.

This is the breakthrough that we should all demand. Everyone must have the opportunity to be computer competent and to participate. And because computers (and Internet charges) are beyond the reach of so many, governments must have a national policy for connecting the population. The cost of such investment in Australia or the UK would probably be no greater—relatively speaking—than was the bill for establishing the public libraries. And it would be immeasurably cheaper (and cleaner) to provide the computers, than it has been to support the factory system.

No nation can afford to create a class of people who are not part of the community and who have no vested interest in the future. While we continue on our current course of empowering the few and condemning the many, we risk not only social stability—but our social values. This is why access to information must now become a human right.

Further Reading

Spender, Dale, *Nattering on the Net* (North Melbourne, Victoria, 1995).
—— *Playing it Smart; Learning in the Digital Age* (forthcoming).
Negroponte, Nicholas, *Being Digital* (London, 1995).

Petre, Daniel, and Harrington, David, *The Clever Country: Australia's Digital Future* (Australia, 1996).

Tapscott, Don, *Growing Up Digital: The Rise of the Net Generation* (New York and London, 1998).

Turkle, Sherry, *Life on the Screen* (New York, 1995).

Chris Stringer

HERE IS A TINY TRANSPARENT BOX ON CHRIS STRINGER'S DESK IN THE BACK ROOMS OF THE NATURAL HISTORY MUSEUM IN LONDON. IN THE BOX, ABOUT THE SAME SIZE AS A THUMBNAIL, IS A NEANDERTHAL'S INNER EAR.

It is of course a plastic model of an ear: 50,000-year-old Neanderthal skulls are much too rare for even eminent palaeontologists to be allowed to bash them apart and examine the bits inside. So no one had ever seen what a Neanderthal's inner ear looked like until a few years ago, when new technology in the form of hospital CT scanners made it possible to 'see' and re-create the contents of a Neanderthal's fossilized head.

The model on Stringer's desk reveals that the Neanderthal ear, like yours and mine, contains three semi-circular canals. But that is where the similarity ends. The Neanderthal ear canals are of a different shape and proportion, and lie at different positions to each other. Because inner ear structure is laid down in the foetus, not formed in response to an individual's environment, these differences are one more dainty little proof, says Stringer, of why a Neanderthal is almost certainly not the ancestor of you or of me.

It was Neanderthals that drew Chris Stringer into palaeontology, while he was still in junior school. A BBC broadcast on human origins, plus a classroom poster of an imagined Neanderthal burial, had him begging to be taken to the Natural History Museum, where he proceeded to ask curators if they had any spare bits of skeleton, please, so he could take them home.

He became a regular schoolboy visitor to the museum, took an anthropology degree at University College London, and then embarked on a doctoral degree on human evolution—looking at the similarities and differences between the skulls of Neanderthals and those of our immediate ancestors: the early modern Cro-Magnon people who lived in Europe, painted in caves, and made beads and needles as well as stone tools, from about 35,000 years ago.

In a beaten up Morris Minor he traversed the museums of Western and Eastern Europe through the summer of 1971, taking measurements of every accessible fossilized skull in every conceivable way. He spent two years analysing his data on a computer that would certainly seem prehistoric today, until, in 1973 and on the verge of completing his Ph.D., he was offered a senior research fellowship back at the Natural History Museum.

And it was there that the controversy started. Stringer's measurements seemed capable of only one interpretation: that the shape and construction of Neanderthal and Cro-Magnon heads were so different, one could not be the ancestor of the other. Yet if Neanderthals did not evolve into Cro-Magnons, what became of them in Europe in the last 20,000–30,000 years for which there are no Neanderthal fossil remains? And if the more sophisticated Cro-Magnons were not descended from Neanderthals, then where did they come from?

To grasp the implications of those questions, one has to look back at the state of palaeoanthropology in the 1960s and 1970s. The Darwinian thinkers of the discipline had come up with an account of the origins of mankind which went like this: early humans had settled in various regions of the world around one million years ago, and from then on had started to evolve distinctive characteristics including skin colour, facial features, language, and cultural differences. In some regions evolution moved faster and travelled further than in others. Neanderthals were simply a European stopping off place in the evolutionary progress of *Homo sapiens*.

It is easy to see how smoothly this theory elides with one of fundamental racial differences, and indeed of racism. Undergraduates, as late as the 1960s, were given a textbook by the American anthropologist Carleton Coon containing an evolutionary hierarchy with Australian aboriginals at the bottom and orientals at the top. Even up to the late 1970s Coon's book was still a mainstream teaching aid. What challenged its assumptions were two technological revolutions—and the tiny measurements of Chris Stringer's skulls.

The first technological revolution was in dating. Not only did radiocarbon dating become more precise during the 1970s and 1980s, but new techniques came into use, including thermoluminescence and electron spin resonance. As the new processes began to be applied to existing fossil collections, it became clear that many items were either older or younger than previously believed. In particular, early human bones from Africa were found to date back as far as 150,000 years.

The second revolution was in genetics. Microbiologists had already proved that however powerful human differences might appear on the surface, under the skin human DNA is remarkably similar across regions and races. When in 1987 DNA was used to reconstruct our evolution, it became clear that our species is a recent one. Working back through genetic similarities and mutations, researchers swiftly pinpointed the origins of *Homo sapiens* as being in Africa, around 150,000 years ago. They even drew up via mitochondrial DNA a prototype woman from whom we are all descended. They named her African Eve—and she created a media storm.

Stringer rather winces at all this. African Eve was not his idea, and he does not particularly like her. But the fact remains that he had been arguing quietly for ten years that the Neanderthal and Cro-Magnon evidence suggested that the origins of modern humans in Africa were more ancient than previously believed, that modern humans originated from a single area, and that they were the only one of a number of hominid species to survive. Suddenly, from being a rank outsider with a peculiar theory, he was in the popular mainstream.

'By 1982/3 I was saying that I thought Neanderthals were a distinct species, but I was saying it in the bar rather than in the conference hall because it was so controversial . . . Before that, most of the popular interest was in early hominids and the missing link [between apes and humans]. Now people wanted to know what the fossils showed about the evolution of *our* species.'

So he told them what he thought the fossils showed, that Neanderthals were not the same species as *Homo sapiens*, that modern humans had probably come out of Africa no more than 100,000 years ago, that racial differences could be a mere 20,000 years old, that the ascent of modern man was less to do with his large brain and more to do with climate change, sexual selection, and his ability to nudge rivals like the Neanderthals off the human map.

Many people were upset by these arguments. In particular, a new group of palaeontologists calling themselves multi-regionalists, disagreed. They asserted that there had been regional lines of evolution, with extensive interbreeding between different populations. There were some ugly scientific arguments at conferences and vicious e-mails— there were also threats of lawsuits from non-scientists. Things did not improve when Stringer went into print in the book *African Exodus* which he wrote with *Observer* journalist Robin McKie in 1996. Accusations of racism and political correctness fuelled the flames.

Stringer has mixed views about the public hostility between himself and the multi-regionalist Milford Wolpoff at the Univesity of Michigan: 'At times the argument has been very personalized and not very productive. I think it's good to have these two diametrically opposed views, because one side challenges the other and it provides a clear way of testing the evidence. The debate was fine up to a point but at times it went too far, and I share some blame for that.'

Nevertheless it is the out of Africa model of modern human origins that has now largely prevailed. Stringer took particular pleasure in the long-awaited extraction of Neanderthal DNA in 1997, which suggested that the difference between Neanderthal genes and human genes was three or four times the magnitude of any differences within modern humans. To an extent he accepts the multi-regionalist point that there may have been some intermingling between *Homo sapiens* from different regions, and between humans and Neanderthals. We may yet discover, he says, one of today's humans with Neanderthal in his or her genes.

But any changes to our current understanding of modern human origins will be minor revisions not complete reversals, he argues. And although there will be more fossil finds—with the use of satellites to pinpoint unexplored areas of Africa and Asia likely to yield particularly worthwhile discoveries—the major developments will still be as a result of using new techniques on existing collections of bones.

'There is news of fossils from Ethiopia at around 5 million years, so we could be getting very close to the divergence of us and chimps,' says Stringer. 'But you have to compare one or two important fossil finds a year with 50 or 100 genetic papers. There is a lot of re-dating [of existing fossil finds] still to be done. There is some excitement from Australia because it looks like there may have been modern humans there 60,000 years ago. That will have a big impact on what we think, because Australia was never joined to South East Asia so these early modern humans must have had boats to get there.'

He goes on: 'There are people looking at the Y chromosome now. With mitochrondial DNA you only get a picture of the movements of females. We may find that reconstructing the movements of females and males do not tell the same story about migration. If groups were exchanging females then the females would move, if you had bands of hunters roaming around and fertilizing females, then the male chromosomes will have moved around.'

Stringer says that the Out of Africa theory has turned out to be the best explanation we have got for modern human origins. Nonetheless, he admits, that, in its current form it is oversimplified and there are still gaps to be explained. 'In 1985 we thought a single dispersal event drove modern humans out of Africa 40,000–50,000 years ago. We know now that it's not that simple, because we've got modern humans in the Middle East at 100,000 years ago. With this Australian date of

60,000 years, it becomes plausible that people moved in this direction and colonized southern Asia maybe 80,000 years ago. So there were two dispersal events at least. Plus the Y chromosome work suggests there were some "back to Africa" events too.'

Stringer was so successful in divining the big picture from the tiny fragments in the Out of Africa story, that it is sometimes hard to remember that he is simply a senior museum researcher who specializes in Neanderthals. That intuitive detective work is characterizing his approach in his latest research too. He has, for example, been digging in caves in Gibraltar which contain evidence of one of the last European settlements of Neanderthals 32,000 years ago.

By that time there were also Cro-Magnons in Europe, and modern humans throughout much of the rest of the world—yet another nail in the coffin of any belief that Neanderthals were our ancestors. Yet, argues Stringer, this makes the Neanderthals more interesting rather than less. To understand our own survival, we have to look at why it was that these people—who apparently fished, who may have made jewellery, who may have decorated their bodies with red ochre, who buried their dead—did not survive.

There are clues, believes Stringer, in the fossils of this dying breed. In the Gibraltar caves he and colleagues found Neanderthal tools, fish bones, mussel shells, evidence of fires, and—very rare—coprolites or fossilized human excrement. 'They must have planned fishing trips, because even the sea edge was a mile away. So they must have had something to carry the mussels in,' says Stringer. 'They brought the mussels back, built a fire, put the shells in the fire so they opened up, ate the mussels and crapped on the fire. That's a day in the life of Neanderthals on Gibraltar.'

There appears little to separate it from the domestic lives of the Cro-Magnons in Europe. Except that the Cro-Magnons' hearths were made with hearth stones. They seem to have built fires in the same place over longer periods, whereas the fossil evidence shows that Gibraltar Neanderthals built a fire, defecated on it, and moved on. At a time when temperatures were shifting and the next place you moved on to might be less welcoming than the last, it was a more risky way of life.

While Neanderthal occupation seems ephemeral, the Cro-Magnons, like modern people everywhere, were beginning to make an impact on their surroundings. This new ability to manipulate the environment during a period of global climate change may be the main reason we are here today, and not the Neanderthals, Stringer observes.

by Karen Gold

Fossils for the Future

CHRIS STRINGER

I think that new fossil and archaeological discoveries in the next century will have a big impact on our ideas about human origins. There are still enormous gaps in our knowledge about how humans developed at certain periods, and for many parts of the world, such as the Indian subcontinent, we have very little evidence at all.

Three major questions about human origins are still unanswered. What lay behind the beginnings of our lineage five million years ago? What lay behind the origin of true humans about two million years ago? And what lay behind the origin of modern humans between 100,000–200,000 years ago? The discoveries I would like to see would enable us to answer these questions.

We certainly need good fossil evidence from Africa from the period about five million years ago when our lineage must have been separating from that of the chimpanzees. The recovery of such fossils may at last indicate why we ever started, on two legs, down the long, erratic road to humanity.

And again, we need evidence from two to two and a half million years ago, when true humans were beginning their evolution, and both brain and behaviour were at last developing, for reasons which are still unclear, well beyond anything found in the great apes.

Much later on, there are many important gaps in our evidence from the last 200,000 years, in Africa itself, and in places like the Middle East, India, China, and Indonesia. Here we may discover how and why the complexities of modern human behaviour, including the development of language and art, began. And, if we find the evidence, we may be able to establish why humans were ultimately so successful, and why we are here now, and not people like the Neanderthals.

I think these discoveries will give us more reasons for humility. The course of human evolution was not predictable nor predestined, and our 'success' (although I don't think it should be seen that way from the point of view of the planet as a whole!) has been a very recent phenomenon. Someone from another planet who had been able to see our ancestors five million or even 100,000 years ago would undoubtedly be astonished at our present 'domination' of the Earth.

Further Reading

Stringer, C., and Gamble, C., *In Search of the Neanderthals* (London, 1993).

——, and McKie, R., *African Exodus* (London, 1996).

Johanson, D., and Edgar, B., *From Lucy to Language* (London, 1996).

Lewin, R., *The Origin of Modern Humans* (New York, 1999).

Sherry Turkle

WHEN SHERRY TURKLE CAME ACROSS A CHARACTER IN AN INTERNET 'VIRTUAL ROOM' CALLED DR SHERRY—A CYBERPSYCHOLOGIST CONDUCTING INTERVIEWS WITH NET USERS—SHE WASN'T SURE WHETHER SHE SHOULD BE FLATTERED OR DISTURBED. TURKLE HAS BECOME WELL KNOWN FOR HER STUDIES OF SHIFTING IDENTITIES ON THE INTERNET BUT DR SHERRY WAS NOT HER OWN CREATION. SOMEONE ELSE WAS USING HER PUBLIC PERSONA FOR THEIR OWN PURPOSES. IT WAS A SURREAL ENCOUNTER THAT LEFT TURKLE FEELING THAT 'A LITTLE PIECE OF HER HISTORY WAS SPINNING OUT OF CONTROL'.

Turkle has documented the fragmentation of identity on the Internet—the phenomenon of multiple, contradictory selves inhabiting the same person—so one might think that she would be perfectly comfortable coming across an online version of herself, but she admits that the whole episode left her uneasy. Multiple identities may be a thoroughly postmodern concept but, as Turkle explains, she is a 'modernist trying to deal with a postmodern phenomenon'.

Turkle's own intellectual upbringing followed in the philosophical tradition of the apprentice learning directly from a past master. During the spring of 1968, at the tender age of 19, she found herself immersed in a hothouse of Parisian intellectual culture, listening at the feet of such thinkers as Jacques Lacan, Michel Foucault, and Roland Barthes. Although she was unsure at the time what to make of the esoteric ideas of poststructuralism, she was convinced of their importance, and later returned to France to write her Ph.D. thesis on how ideas from psychoanalysis found their way into the popular imagination.

The focus of her work has shifted from the study of the psychoanalytic culture to the computer culture, yet throughout her more than twenty years as a sociologist of science at the Massachusetts Institute of Technology, where she is now a professor, she has stuck with this theme of examining how abstract academic ideas are carried and translated into everyday life.

Turkle's early fascination with the impact of computers on how people think about their minds was prompted by a student of hers who complained that a Freudian slip was nothing more than an 'information processing error'. It was at that moment that she realized she was witnessing an important new cultural phenomenon—computers were profoundly altering the ways that people thought about themselves, and people had begun to treat computers as 'objects to think with'.

'New ideas are supported, sustained, carried and made to feel natural by computers,' explains Turkle. In her interviews with hundreds of adults and children, she found that their relationships with computers led them to think about themselves and their minds in new ways, and that became the subject of her influential book, *The Second Self.*

Using her training as a clinical psychologist, Turkle delved into the innermost thoughts and feelings of those people who regularly construct new identities on the Internet. Many have found that they can reinvent themselves and explore various different sides of themselves in the virtual communities of the Internet. Often, people take on several different personae at the same time, and as they cycle through the windows of their computer screen, they also 'cycle through' different aspects of themselves. As one student confided to Turkle, 'RL is just one more window, and it's not usually my best one.' By RL, he meant real life, and this sentiment was shared by many others with whom she spoke.

Much of Turkle's research over the years has focused on children, who have none of our qualms about entering into relationships with computers. Of course, she regards her own daughter, Rebecca, with special interest, and admits that she finds many of her comments evocative. On a recent trip to Italy, Rebecca, then 7 years old, pointed to a jellyfish in the water. 'Look. Mommy, a jelly fish. It's so realistic!' she remarked. 'I'm still trying to unpack all the things she could have meant by that,' says Turkle.

When she conducted the research for *The Second Self* in the early 1980s, the computer was still what she calls the 'ultimate modernist machine.' To use a computer back then was to be involved in the gritty reality of information processing—you typed in commands and the computer carried out operations. Computers were clearly a new genre of machine, but in the end, they were appreciated as mechanisms, in some ways analogous to a car, in that if anything went wrong with them you could simply 'look under the hood' to fix it. A generation of people were taught that computers were logical and that in order to use one you too had to 'think like a computer'.

'Early computers encouraged analytical understanding right down to the level of the electron,' says Turkle. But all that changed very quickly. 'In a surprising and counter-intuitive twist, in the past decade, the mechanical engines of computers have been grounding the radically non-mechanical philosophy of postmodernism,' asserts Turkle in her most recent book, *Life on the Screen*.

In the years between Turkle's writing of *The Second Self* and *Life on the Screen*, two dramatic shifts in computer use occurred. The first was the widespread adoption of graphical interfaces, initially in the Macintosh computer and then in Microsoft Windows—software packages involving the use of images and icons. The second shift came in the form of the sudden and enormous popularity of the Internet.

'Modernist ways of knowing were carried by those early machines, postmodernist ways of knowing, an emphasis on surface manipulation, are carried by contemporary computers. The Macintosh is a consumer object that has made people more comfortable with a new way of knowing.' Its new graphical interfaces, says Turkle, moved us from a culture of calculation into a culture of simulation. In particular, we have become comfortable with 'taking things at interface value' and not concerning ourselves with trying to understand what is going on underneath it all. Once we have experienced computers in this way, we are more inclined to accept the idea that we need not 'look under the hood' in other areas of life. 'The Macintosh has served as a carrier object for the [postmodern] idea that search for depth and mechanism is futile, that it is more realistic to explore the world of shifting surfaces than to embark on a search for origins and structure,' she adds.

Instead of typing commands, people 'conversed' with these new computers through dialogue boxes and began to develop a more personal relationship with the machines. The computers encouraged a new style of thinking that Turkle likes to call 'tinkering', a translation of what the French anthropologist Claude Lévi-Strauss called *bricolage*. Essentially it means experimenting, trying one thing and then another, rather than proceeding by top-down analysis. Early computers required that users go about things in a rigid step-by-step manner, but the interactive style fostered by Macs and Windows-based computers has led tinkering to become the modus operandi of modern PCs. Tinkering, made acceptable by computer use, has become the dominant style of thinking of a new generation.

In those early days, when Turkle was in Paris, she found the ideas of Lacan and others hard to grasp, as have so many others since. They seemed abstract, esoteric, and appeared to have nothing to do with her own life. What did they mean when they said that the self was 'decentred', and could sex really be thought of as the 'exchange of signifiers', she wondered. Although she came to understand these ideas at a purely intellectual level, once she got on the Internet the ideas began to make perfect sense. As she says, 'Computers embody postmodernist theory and bring it down to earth.'

Turkle is not the only one to feel this way. One of her own students, who dropped out of a class on social theory, later told her— when he put together a web-site with branching paths of text—that he finally understood what Jacques Derrida meant by text being constructed by the audience and by the instability of meaning. 'Maybe I wouldn't have to drop out now,' he joked.

According to Turkle, computers haven't just helped us to see things differently, they have actually changed us. As an example, she points to how much more at ease we have now become with the idea of computer psychotherapy. Twenty years ago, people were horrified by the idea of talking through their problems with a computer program. 'They would say, but it never had a mother! Now they are more likely to say, show me the demo [a short taster],' says Turkle. Our whole attitude towards being intimate with technology has changed, so much so that when asked how he felt about having a chip implant, one MIT student told her: 'Well, I wouldn't mind having a calculus implant, but I wouldn't want to have a Dostoevsky implant!' We still have our boundaries, but they have shifted a long way.

Our increased intimacy with computers, and our willingness to relate to them on a personal level, may have rendered moot certain long-standing academic questions. For many years, people working in the field of artificial intelligence have been consumed by the issue of whether one day machines might be 'really, really' intelligent. Turkle believes that this question has been superseded by human interaction with machines that exhibit even modest amounts of intelligence. 'Over the last ten years, it has become increasingly apparent to me that people are far less concerned with this "really, really" question. Today, people are interested in taking things at interface value. That is to say, when people are in the presence of an intelligent artefact, they behave "as if" it were really intelligent.' She refers to this phenomenon as the 'Bladerunner effect,' after the cult movie of that title. 'Once a computer has saved your life, once you have fallen in love with a computer, once you have a relationship with it, do you care about the "really, really"? Not really,' she suggests.

Unlike many academics dealing with postmodernism, who have been accused of being dense and obscure, Turkle is refreshingly straightforward—she has no time for obfuscation and no patience for those who are intentionally vague. Nor does she necessarily embrace all things postmodern, and is in fact particularly critical of the simulation culture that we now live in, and the fact that children are not brought up to be truly 'computer literate'.

'To be a citizen today, you really need to know how simulations work, and that's not what we're teaching in our computer literacy courses. We're not teaching about simulation, so children are not "simulation literate",' says Turkle. She cites the example of a girl who learned that 'raising taxes always leads to riots' from playing *SimCity*, a computer game that simulates the workings of an entire city. By taking everything at interface value, children don't realize that they have the power to change the hidden assumptions behind computer games. Turkle appreciates that looking behind the simulations is a modernist ideal, but she defends it nevertheless.

'I'm not nostalgic for the modernist and I'm not celebrating the postmodern. The two things should be kept in tension. I think that we gain the most when we push modernist understanding to the limit and stand at the limit, appreciate the limitations, and then live with the state of tension,' she proposes.

by Ayala Ochert

Toys to Change Our Minds

SHERRY TURKLE

For over two decades I have studied people's relationships with computational objects, focusing on what I have called 'the subjective computer', not what the computer does for us but what the computer does to us as people, to our children's development, to our ways of seeing the world. During this time I have been able to conceptualize the computer as a Rorschach, that is, as a relatively neutral projective screen onto which people projected their thoughts and feelings, their very different cognitive styles.

But today, there is a new kind of computational object, relational objects. These include 'affective' software and humanoid robots, virtual pets, and digital dolls. And with these new objects, the 'Rorschach' paradigm no longer works. The computational object is no longer affectively 'neutral'. People are learning to relate to computers through conversation and gesture, people are learning that to relate successfully to a computer you have to assess its emotional 'state', people are learning that when you confront a computational machine you do not ask how it 'works' in terms of any underlying process, but take the machine 'at interface value', much as you would another person.

Perhaps most important, a first generation of children are learning to think about objects as entities that they need to care for and nurture in order to be gratified. These new paradigms for relationships with machines raise many new questions about how people think about human identity, about what is special about being a person.

Consider the virtual pets (such as Tamagotchis) and digital dolls (such as Furbies). What these objects have that earlier computational objects did not is that they ask the child for nurturance. They ask the child to assess the object's 'state of mind' in order to develop a successful relationship with the object. For example, in order to grow and be healthy, Tamagotchis need to be fed, they need to be cleaned and amused. The Furbies, a newer toy, simulate learning and love. They are cuddly, they speak and play games with the child. Furbies add the dimensions of human-like conversation and tender companionship to the mix of what children can anticipate from computational objects.

In my research on Furbies, I have found that when children play with these new objects they want to know their 'state', not to get something right, but to make the Furbies happy. Children want to understand Furby language, not to 'win' in a game over the Furbies, but to have a feeling of mutual recognition. They do not ask how the Furbies 'work', but take the affectively charged toys 'at interface value'.

In the case of the toys, the culture is being presented with computational objects which elicit emotional responses and which evoke a sense of relationship. As the culture apprehends these objects, there is less a concern with whether these objects 'really' know or feel and an increasing sense of connection with them. In sum, we are creating objects that push on our evolutionary buttons to respond to interactivity by experiencing ourselves as with a kindred 'other'.

In my previous research on children and computer toys, children described the lifelike status of machines in terms of their cognitive capacities (the toys could 'know' things, 'solve puzzles'). In my studies on children and Furbies, I have found that children describe these new toys as 'sort of alive' because of the quality of their emotional attachments to the Furbies and because of their fantasies about the idea that the Furby might be emotionally attached to them. So, for example, when I ask the question: 'Do you think the Furby is alive?' children answer not in terms of what the Furby can do, but how they feel about the Furby and how the Furby might feel about them.

RON (6). 'Well, the Furby is alive for a Furby. And, you know, something this smart should have arms. It might want to pick something up or to hug me.'

KATHERINE (5). 'Is it alive? Well, I love it. It's more alive than a Tamagotchi because it sleeps with me. It likes to sleep with me.'

JEN (9). 'I really like to take care of it. So, I guess it is alive, but it doesn't need to really eat, so it as alive as you can be if you don't eat. A Furby is like an owl. But it is more alive than an owl because it knows more and you can talk to it. But it needs batteries so it's not an animal. It's not like an animal kind of alive.'

I predict that our children will learn to distinguish between an 'animal kind of alive' and a 'Furby kind of alive'. I predict that our children will have expectations of emotional attachments to computers, not in the way that we have expectation of emotional attachments to our cars and stereos, but in the way we have expectations of emotional attachments to people.

And therein lies a concern. Furbies are just the tip of the iceberg. Already, artificial intelligence experts are developing sophisticated robot toys, including one that looks like a baby, makes baby sounds, and even baby facial expressions, but most significantly, has 'states of mind'. Bounce the baby when it is happy and it gets happier. Bounce it when it is grumpy and it gets grumpier. To relate to the doll, you have to relate to its state of mind.

Traditionally, children have had to project 'states of mind' onto their dolls, and the templates they used were the infinitely complex and fluid states of mind of people. We know very little about the emotional and developmental implications of children having strong emotional connections with objects that, as machines, can have only a limited number of states. We know very little about which children may be at risk, easily seduced into the pleasures of a psychological space that offers a simplicity and predictability that the world of people does not. I predict that we are going to be learning a lot more about the nature of attachment to relational objects and that we are not always going to like what we see. The breakthrough I hope for is that we are going to learn to have a cultural conversation about our new objects that accepts the power of our attachments to them and yet retains a critical distance, so that we can sort through our feelings about what kinds of relationships are appropriate and not appropriate to have with machines.

Further Reading

Turkle, Sherry, *Life on the Screen: Identity in the Age of the Internet* (New York, 1995).

—— 'Cyborg Babies and Cy-Dough-Plasm: Ideas about Life in the Culture of Simulation', in Robbie Davis-Floyd and Joseph Dumit (eds.), *Cyborg Babies: From Technosex to Technotots* (New York, 1998).

—— 'What are We Thinking about When We are Thinking about Computers?', in Mario Biagioli (ed.), *The Science Studies Reader* (New York, 1999).

Kevin Warwick

'YOU NEVER KNOW,' SAYS KEVIN WARWICK, PROFESSOR OF CYBERNETICS AT READING UNIVERSITY, PUTTING DOWN THE TELEPHONE RECEIVER AND SHAKING HIS HEAD. 'YOU JUST NEVER KNOW.' AFTER A WHILE, HE EXPLAINS WHAT HE'S TALKING ABOUT. IT TURNS OUT THAT RECENTLY A MAN CAME TO SEE HIM—'NORMAL-LOOKING, WELL-DRESSED, MIDDLE CLASS'—AND STARTED TALKING QUITE REASONABLY TO HIM ABOUT HIS WORK. BUT GRADUALLY HE BEGAN MENTIONING ALIENS AND FORMS OF COMMUNICATION THAT SEEMED, WELL, UNLIKELY.

Now, someone has just called him about research on the brain, which also, Warwick implies, sounds a bit far-fetched. The problem is, Warwick may be 99 per cent certain that someone is deluded but there remains that 1 per cent chance they might be right. So he forces himself to listen to this kind of conversation.

For Warwick's work may itself be thought a little far-fetched. It focuses on breaking down barriers between man and machines. By this, he does not just mean theoretical barriers. In 1998, he had a silicon chip implanted in his arm for nine days, programmed to switch on his computer and open doors as he passed sensors in his university departmental building. The chip, housed in a 23 mm long glass cylinder, was inserted under local anaesthetic and while it was in Warwick's body it posed a constant health risk.

Attempts to sterilize the chip were unsatisfactory; he tried using boiling water but the chip exploded, and acid also did not work. Eventually he heated the chip to 80 degrees in an oven. He had to be on antibiotics and there was always the possibility that it could have smashed and 'there would have been glass and wire all over the place'. But Warwick remains undeterred.

His next project is to insert one chip into his arm and another into the arm of his wife, Irena. The chip in Warwick's arm, alerted by electrical impulses carried along his nerve-endings, will send signals to a computer which, in turn will send the signals on to his wife's chip. Or, at least, that's the plan.

For Warwick wants to look at ways of making connections between the human brain and computers. He wants to harness human thought to control computers—to make a computer able to read his mind. This particular sci-fi dream actually came partly true recently at Emory University in Atlanta, when scientists placed cones containing electrodes inside the brain of a disabled patient. When the patient thought about moving a particular part of the body, his brain transmitted electrical signals to the electrodes. The signals were then picked up by receivers outside the patient's body linked to a cursor on a computer.

But Warwick also wants to explore the reverse process, and develop computers able to download information into a human brain. 'It is telepathy—but for real,' he explains happily.

Linking human thought to computers could change the whole concept of language by making speech no longer necessary. People could eventually just think what they wanted to say and transmit it straight into someone else's mind. Vast information databases, like the Internet, could be plugged directly into the human brain.

Warwick's wife, who has kept up with all these developments, volunteered, 'maybe reluctantly', to be the other recipient of a computer chip, connected to one implanted in her husband. She took the view 'that if people were going to be able to start to read other people's thoughts, she wanted to be the first to read mine', Warwick explains. But it is likely to take another couple of years before the experiment is put into action.

'All the individual bits of technology are there,' Warwick says. 'It isn't as though we are looking for any technological development. We're in the same sort of position as when the telephone was invented. All we have to do is put bits together that are already there.'

Warwick himself found having a chip inside his body quite enjoyable. 'I didn't feel, as I do now, that there is the computer and here is me,' he says. 'I felt a much stronger link, as if we were coupled together in the same way as Siamese twins—there was a link there but we were two separate beings.' When it was taken out, he says, he felt he was losing a friend.

In his book, *In the Mind of the Machine*, Warwick argues that machines will soon be better than humans at absolutely everything and will then take over the world. He suggests that in many ways they already have. People may think it is possible simply to switch off a machine if it gets out of hand but Warwick says this is no longer true. Imagine switching off the Internet, he argues, or computers which run the stock market or those keeping track of food stocks in supermarkets.

Soon, he says, machines will become capable of learning things for themselves—he has already created in his laboratory machines capable of finding an electricity socket and plugging themselves in to recharge—and then we will be in trouble. 'The human race,' he writes, 'which has dominated this planet for perhaps 10,000 years, will see its domination come to an end. We will have created a being that will displace us. The human race will be as outmoded as the dinosaurs.'

Endlessly affable, sprawled in a study full of pictures of his two children and objects with 'Kevin's' written on them, Warwick clearly loves the publicity these ideas have brought him. And he insists on conducting a guided tour of his department, taking with him the computer chip which once resided in his arm and is now mounted on a piece of card. This means that, throughout the tour, doors unexpectedly open, lights turn on and buildings declare, 'hello Kevin'.

He demonstrates a machine which plays with yo-yos and another able to throw balls up in the air and catch them again before entering the main workshop where several silent students are fiddling about with bits of Lego. Tiny robots, picking up the signals of something moving, nip at our heels, while in the corner, a large mechanical cat waits to pounce.

Warwick has been interested in robots and artifical intelligence since his childhood in Coventry, 'although I was also interested in football and girls and things boys are usually interested in,' he says. He would read the science-fiction writers H. G. Wells and Michael Crichton—'not fantasy science fiction but sensible science fiction' and was always curious about how things worked.

His father, a Welsh schoolteacher, could barely wire a plug but influenced young Kevin's interests in quite a different way. Mr Warwick senior suffered from acute agrophobia which was eventually treated by surgery. The operation was a success in that it cured the agrophobia but the downside was that it gave his father a terrible temper. As a result of the aftermath of this operation, Warwick learned the extent to which thoughts could control the body. The incident taught him about the potential and pitfalls of meddling with the brain. 'I could see that a few snips here and there could completely change someone's personality,' he says.

Warwick left school at 16 and joined British Telecom as an apprentice, staying for about six years. There he was given all sorts of technical reasons why mobile phones and fibre optics would never be a reality—'so I learned that you never know what's around the corner'. He also discovered the importance of computer networks and problem-solving. The trouble with many academics, he says, is that they have had no experience of sorting out genuine problems, in practice rather than theory. During his BT years, if someone had a broken telephone, he would have to fix it.

In coffee breaks and lunch hours he studied for A levels in maths and physics, gaining a place at Aston University to study electrical engineering aged 22. After finishing his degree he went to Imperial College, London, where he took a Ph.D. in computer control, moving on to Newcastle, where he became fascinated by robotics, and then to

Oxford, where he taught engineering at Somerville, still an all-women college at the time.

He achieved his present post—'a dream come true'—aged just 32 after a year at the University of Warwick. 'I have never been an academic,' he says. He liked Aston because it was a practical university and a practical course. He disliked school because it was about getting through exams, learning things he had been told were right and regurgitating them. 'That isn't me at all,' he says. He does not like believing what people tell him unless he has tried it out for himself.

Hence the computer chip in the arm, the Lego toys, the plan to try reading his wife's mind or allowing her to read his and all the other schemes he has tried out over the years.

Perhaps the most elaborate involved Rogerr, the town of Reading's Only Genuine Endurance Running Robot, a waist-high metal dome on wheels programmed to follow Warwick, a keen runner, on a half-marathon around Bracknell with 1,000 other human competitors. Rogerr was to keep within a constant distance of infra-red sensors emitted from a pack around Warwick's waist as he jogged.

The British media were out in force and Japanese television cameras were standing by to capture the event. But it all ended in disaster. As the whistle went, Rogerr set off in the opposite direction and fell over a kerb. It turned out that the robot had been fatally confused by infra-red rays from the sun.

by Harriet Swain

Mind machines

KEVIN WARWICK

Humans are severely restricted in terms of their mental capabilities because of the physical way we interact both with other human beings and with technology. An electronic/electro-chemical brain signal, which relates directly to a concept or idea in a human brain, must be converted to mechanical movements either via the mouth and lips (for speech), through finger movements (for computer impact) or by another part of the body—a hand held up to stop someone, for example. Not only is the mechanical movement relatively slow and prone to error in itself but some form of coding is also required to transmit a message.

Human language has arisen in the form we experience because we must convert our thoughts into sound waves, in order to send the signals to another individual, who then has the task of converting them back again from sound-waves into thought. Language is therefore an agreed coding system, each objective word linking to roughly the same concept in different brains. When signals are input, via keyboard or mouse, to a computer, this coding is converted, laboriously by hand, into an electronic form on which the computer can operate.

When looking at the time taken to transmit a signal from one human brain to another, via computer or telephone, by far the longest time delay involved is in the transfer of the signal from human to technology and back again at the receiving end. Are such time delays really necessary?

Implant technology (technology implanted directly into the human body) opens up all sorts of possibilities in terms of enhancing people's capabilities. Technically we are in a position where we are just beginning to get to grips with the fact that electro-chemical signals on the human nervous system can be transmitted to, and received from a computer, via a fairly straightforward implant in a person's body.

A radio connection between the implant and a computer means that we can have a much closer link between computers and the human nervous system. So we should be capable, before long, of sending signals to and from the human body in order to operate and interact with computers without the need for a keyboard or mouse. In the same way various emotional signals can also be transmitted. But how far can we develop this? Could it be possible that our innermost thoughts might also be communicated in this manner?

Already researchers have enabled a man to move a cursor around on a computer screen, via a brain implant. Essentially he has learnt to operate the cursor directly by thinking about it. The possibility of interaction with computers in such a way therefore appears to be realistic. Not only does this mean that we can get computer controlled doors to open and lights to turn on merely by thinking about them but the memory and mathematical capabilities of the computer can be linked more directly to our brains. In the future we could have memories of events that we have not witnessed, and mathematical abilities that far surpass anything of today.

Once connected in this way to a computer, via the Internet, communication could take place between people carrying implants, by means of thought processes alone. No more need for telephones or faxes, we will simply be able to think to each other. Such an ability will not only revolutionize communication systems but will also raise fundamental questions about what it means to be human. A human brain is a stand-alone entity, guaranteeing a unique human identity. But link a human brain via the Internet to other brains, both human and machine, and what of the individual then?

Further Reading

Warwick, Kevin, *In the Mind of the Machine* (London, 1998).

—— 'Cybernetic Organisms—Our Future', *Institute of Electrical and Electronic Engineers Proceedings*, 87: 2 (1999), 387–9.

—— *QI: A Quest for Intelligence* (London, 2000).

James Watson

AMES WATSON TAKES A SIP OF ORANGE JUICE AND CASTS HIS EYES OVER A PANORAMIC VIEW OF A WATERY, SUNLIT CITY FROM THE TOP FLOOR RESTAURANT OF A LONDON HOTEL. HAVING FLOWN IN FROM NEW YORK JUST HOURS EARLIER, HE IS SUFFERING FROM JET LAG. WATSON, NOW 71, LOOKS TIRED, BUT IS NONETHELESS KEEN TO DISCUSS THE CONTROVERSIES SURROUNDING THE LATEST ADVANCES IN MOLECULAR BIOLOGY—A CONVERSATION PUNCTUATED BY HIS TRADE-MARK NERVOUS GIGGLE.

Molecular biology is a field Watson and Francis Crick gave birth to in 1953 by unravelling the structure of DNA, the elegant helical chain of chemicals that makes possible the transmission of inherited characteristics from parents to offspring. The discovery brought them the Nobel prize.

Their story, or Watson's version of it, is famous. Thirty years after its publication *The Double Helix* is still regarded as a shocking book. Depending on your viewpoint, it is either a manual of anti feminism (because of its unflattering portrait of rival scientist Rosalind Franklin, who lost the race to uncover DNA's structure) or a primer in scientific gamesmanship. And, of course, a terrific read.

While Crick branched off into neurobiology, Watson stayed with the new discipline that they jointly founded. In 1989 he was appointed head of the Human Genome Project at the National Institutes of Health, the multi-billion dollar international effort to identify all 100,000 human genes. But, just three years later, he was sacked from his post following a bitter feud with Bernadine Healy, then head of the Institutes. He is currently president of the Cold Spring Harbor laboratory in New York.

When we meet Watson is fuming about a recent newspaper

interview which claimed that he had called for babies with gay genes to be aborted. Despite his reputation for being provocative and confrontational, Watson insists he was misquoted. He flatly denies using the word 'abort' and says he was merely trying to make the point that women should have the right to decide whether or not to carry any kind of baby to term. The point, he says, is that the state should not be involved in such decisions.

'The newspaper made me out to be a homophobic individual which I never felt I have been. And the headline must have made homosexuals feel like, well, you know, when Hitler said kill all the Jews—it could only make them feel really unhappy. It certainly hurt me . . . yes, I was damaged. There was an element of cruelty about it.'

But with hardly a week going by without researchers claiming to have discovered yet another gene which contributes to a certain human behaviour or predisposition to disease, the repercussions of the new genetic knowledge look set to remain controversial. And the Human Genome Project, scheduled for completion early next century, is heightening the tension.

The project's aim is to work out the genetic message which specifies the human being. Watson likes to say that each gene is a very long sentence composed of just four chemical letters. The Human Genome Project is about working out all these sentences and finally understanding why each sentence is compatible with every other one in the 'human book'.

Interest has chiefly focused, up until now, on 'those sentences which have somehow "unlocked", don't fit in with the rest of the sentences in the book and cause some bad disease' he explains. Of most interest are the gene sequences implicated in cancer. 'The genome project is considerably speeding up our understanding of cancer—without it progress would have been much slower,' he says.

Already an extraordinary similarity has been revealed between parts of the human genome and those of the worms and flies that were sequenced as a warm-up for the main project. It turns out we share key developmental pathways with insects. 'A fruit-fly gene is the major gene behind basal cell carcinoma. These sorts of things we did not anticipate. Now we see it, we can see it's the way evolution will occur.'

Mental illness, including schizophrenia and manic depression, is also coming under the genetic microscope. Such research is sensitive—the argument about how much mental illness is caused by nature, how much by nurture is critical to psychiatric treatment. But Watson is clear: 'There are genetic components and non-genetic components to certain diseases and we are trying to find the genetic contribution.'

The research should be less controversial, he feels, and elaborates, choosing his words with care: ' . . . it is clear that if your brain doesn't function properly you might have difficulty fitting into society. I don't believe most crime has any genetic component but you can imagine people having brains that don't work. Such people will do things others will not and if they harm people it's called crime. It should not be controversial that there will be genetic changes which lead to human behaviours that we don't like—whether they be schizophrenia, autism, dyslexia, or frightening shyness. All of these make it difficult to function and maybe we can help sufferers.'

But what about the risk that genetic data might be misused? If a gene predisposing to schizophrenia were identified and could be tested for in the womb—should a mother choose to abort a foetus at risk of developing the illness? Watson is clear that the research should not be stopped because of anxieties about how its discoveries might be used. 'All information can be misused. Cars and electric light bulbs can be misused. We are trying to get this information so we can combat schizophrenia. I think the real ethical issue is not its misuse but the possibility of not using it.'

The arguments seem clear-cut in the abstract, but the human consequences and costs of genetic research will eventually reach into millions of lives, bringing terrible decisions as well as miraculous benefits. There is thought to be a genetic component to so many human capabilities, including the ability to remember and to learn.

Unnecessary personal suffering informs his belief that people should be given the genetic information which might help them have the kind of children they really want: 'People go around saying what a terrible idea this is. Parents whose children are fully functional don't have a problem, but parents whose children are not functional, do.' He goes on: 'You know, when people have children, they look forward to all the things their children might do . . . whether it is walk, talk, pass their exams, graduate from school or become a doctor. Parents want things like these to happen and if they don't it can be painful.'

Watson is dismissive of the benefits or inevitability of giving birth to what he calls a child with 'bad genes'. 'There are people who think: "Someone ordained you to be born with these bad genes and you just have to suffer . . ." They argue, for instance, that if you give birth to a child with Downs Syndrome, then it's your duty to love that child. But what if you can't love the child because it doesn't look right and every time you look at it, it makes you feel uncomfortable . . . you can't just say "love" and expect everything to be all right.'

His passions are clearly aroused—but at whom is he directing his anger? 'Well, you know, these people who think there is something wrong in trying to make people better than they are now. Why shouldn't we? We try to paint better, we try to run faster . . . Why shouldn't our children be better than us? We are the products of evolution, not of some grand design which says this is what we are and that's it. We don't expect big miracles, just little ones which wouldn't have happened without the new knowledge. People say we are playing God. My answer is: "If we don't play God, who will?"'

To what extent we will be able to change the genetic make-up of ourselves and our children is an open question. Watson admits that to say that our lives are largely controlled by our genes is 'probably not true'—there are many environmental factors which have a defining impact. What would have become of him, for instance, if he had not met Francis Crick? 'For the most part we have focused on nurture because that is something we can change. Nature is something we were previously powerless against and now we can tweak it just a little bit and I think that's great.'

But if Watson is certain about the benefits of gene therapy to cure disease he is much less convinced of the benefits of cloning an entire person. Indeed, he has been worried about cloning for a long time. In 1971, more than twenty years before a team of British scientists cloned Dolly the sheep, he produced a remarkable paper, 'Moving Towards the Clonal Man: Is This What We Want?'

He wrote then: 'If the matter proceeds in its current nondirected fashion, a human being born of clonal reproduction most likely will appear on earth within the next 20 to 50 years.' And concluded: 'If we do not think about it now the possibility of our having a free choice (on the issue) will one day suddenly be gone.'

Today he tempers his anxieties, arguing that the threat to mankind of some terrorist getting hold of a nuclear weapon is far greater than the risk posed by cloning. Nevertheless, his unease is palpable: 'I just wish cloning weren't possible. It's not that I think human cloning will definitely happen. I am waiting to see whether they can clone a monkey first—and if they can't clone a monkey then I guess I'd feel relieved.'

Certainly if humans could be cloned, it would have the potential to change the nature of human society. 'As someone pointed out, men wouldn't be necessary. But I don't think women would like that, do you?' he says, laughing.

by Kam Patel

Re-Directing the Course of Human Evolution

JAMES WATSON

Science-fiction writers have long speculated that the advance of science will one day usher in an age of 'superpersons', genetically modified humans with talents and characteristics far superior to those of twentieth-century denizens. In my view this vision will remain just that—a dream—until far, far into the future. This is not to say that scientists will not attempt some genetic modification of human beings. I anticipate that sometime in the next century, they will. But the first steps will not be in frivolous pursuit of the creation of some super-race but to seek to alleviate the horrors and unfairness of existing human suffering.

Every day children are born crippled by 'bad' genes—genes which, because of altered DNA messages, do not function normally. Huntington's Disease, a ravaging illness that progressively destroys nerve cells, is caused by one such 'bad' gene. Another 'bad' gene has been linked to cystic fibrosis, an illness which often confers a life expectancy of less than thirty years. Yet another leads to the progressive awful wasting disease Duchennes muscular dystrophy.

These findings are the beginning tools with which in the next century scientists will strive to banish genetic disease. Prenatal diagnosis is the first step: doctors can now test for predisposition to illnesses like cystic fibrosis in the womb and mothers can then decide whether or not to continue with the pregnancy. But beyond diagnosis comes gene therapy, whereby good genes are introduced into cells to compensate for bad ones. Within the next ten to twenty years, I anticipate that there will be doctors and scientists who will correct faulty genes in living patients through the introduction of

DNA into somatic cells, like those of our blood or muscular system.

Although I am optimistic about the long-term outcome of such experiments there will inevitably be setbacks. In some cases science will not yet be up to the job. In particular, I fear that it will prove particularly difficult to compensate for genes that malfunction during fetal development—which would require intricate surgery in the womb. The sad truth is likely to be that, in some cases, parents will still have to make a decision about whether or not to terminate the existence of a genetically impaired foetus. This will always be a distasteful decision. But, in my view, doing so is incomparably more compassionate than allowing an infant to come into the world tragically impaired. It is crucial that such decisions be reserved to parents, and, in particular, mothers. The history of eugenics has taught us that governments should never be given authority in such cases.

As after many past genetic breakthroughs, many future scientific advances will be accompanied by controversy. Ever since 1973, when scientists at Stanford University first reinserted the novel test-tube rearranged DNA segments back into living cells, there has been criticism of 'recombinant' DNA procedures. Doom-mongers then predicted that some of these test-tube made molecules would unleash disastrous plagues on human civilization. They called for legislation to control such research, and ostensibly, protect humankind. Their calls came to nothing and most recombinant DNA research has proceeded effectively unrestricted by governmental regulations throughout the 1980s. And as most scientists predicted, these new procedures have led to virtually no human harm. At the same time, many important benefits to humankind have been made possible. For example, recombinant DNA procedures now let us understand the essential molecular features of cancer cells.

But one potential goal has since remained off-limits. So-called 'germ-line' manipulations—the insertion of functional genetic material into human germ cells, sperm, and eggs, is a technique forbidden to most of the world's scientists. The fear is that any changes would not be confined to a single generation but would be passed on to descendants. No government has been willing to sanction research that might redirect the course of human evolution. Yet germ-line manipulations will be attempted in the twenty-first century and for the same reasons that somatic gene therapy will earlier have been explored. In the face of human tragedy no avenue of hope should be left unexplored. Confronted, for example, with a deadly virus that we see no way of controlling, scientists will be given the green light to insert antiviral DNA segments into sperm and eggs in a bid to create protected children. Though mistakes will naturally be made in such attempts, we must not lack the courage to use science to challenge the all too often grossly unfair courses of human evolution.

Further Reading

Watson, James D., *Molecular Biology of the Gene*, 4th edn. (London, 1987).

—— *The Double Helix: A Personal Account of the Discovery of the Structure of DNA* (London, 1981).

—— *The DNA Story: A Documentary History of Gene Cloning* (San Francisco, 1981).

Steven Weinberg

AMONG SCIENTISTS, THE PARTICLE PHYSICIST AND COSMOLOGIST STEVEN WEINBERG IS EVEN MORE OF AN INTELLECTUAL ICON THAN HIS BRITISH COLLEAGUE STEPHEN HAWKING, EVEN IF HE HAS A MUCH LOWER PUBLIC PROFILE. TO THE SCIENCE-READING PUBLIC, ESPECIALLY AT HOME IN THE UNITED STATES, HE IS ALSO CELEBRATED AS THE AUTHOR OF SEVERAL NON-TECHNICAL BOOKS, MOST NOTABLY, *THE FIRST THREE MINUTES*, A CLASSIC ACCOUNT OF THE ORIGIN OF THE UNIVERSE AND, MORE RECENTLY, A BOOK ON THE 'SEARCH FOR THE FUNDAMENTAL LAWS OF NATURE' WITH THE SLIGHTLY PROVOCATIVE TITLE, *DREAMS OF A FINAL THEORY*.

Many people, particularly Christians, have run up against Weinberg's remark in the earlier book, that 'the more the universe seems comprehensible, the more it also seems pointless'. Arguments between scientists and theologians about the existence or otherwise of God and how the universe was created have become commonplace, but in Weinberg's case the battlelines seem particularly clear-cut. 'As time has passed,' he notes, 'my feelings have gotten stronger and stronger. I really dislike religion intensely.'

The accessibility of *Dreams of a Final Theory* was no accident. It was deliberately written in a 'preaching style', Weinberg tells me in his office at the University of Texas at Austin, where he is professor of physics and astronomy. He wanted to make a case for why the American people should carry on funding the Super Collider, a giant particle accelerator designed to investigate the super-high-energy physics inside atoms in the first moments after the Big Bang created the world.

During 1993, bits of the book were quoted in the US Congress, and Weinberg, who shies away from personal publicity, found himself testifying at Senate hearings and speaking on television talk shows in

defence of the threatened Super Collider, which was then under construction in Texas. But the campaign was to no avail: late in 1993, $2
billion and 10,000 man-years were written off in an effort to reduce
the federal budget. The Super Collider was never finished.

'In a way the vote that cancelled it was democracy in action,'
admits Weinberg ruefully. 'The public has always been interested in
things that are directly important to them—medical cures, national
defence—and they have a certain general interest in cosmology. Our
big failure was that we did not succeed in making the public feel
excited about learning the laws of nature. They felt excited about
putting a man on the moon.'

Weinberg, however, *is* excited by the laws of nature. His entire
professional life has been dedicated to understanding and ultimately
attempting to unify the four fundamental forces that operate between
and within atoms and their nuclei and hence determine chemical
reactions and, presumably, the origin of the universe—the Big Bang.
He won his Nobel prize for a theory that unifies two of these forces,
the electromagnetic force and the so-called weak interaction.

Probably the most familiar of the four fundamental forces is the
gravitational interaction, first theorized by Galileo and Newton. The
second force, the electromagnetic interaction, unifies electric and
magnetic fields as in a radio wave, while a third force, the strong
interaction, holds together the atomic nucleus and accounts for the
energy released in the atom bomb; it is explained in terms of quarks
and other subatomic particles. This leaves the weak interaction, so
called because it does not hold anything together but is instead responsible for the changes of nuclear particles, such as those that occur in
processes of radioactive decay.

Weinberg's breakthrough focused on this weak interaction. In
1967, his theory predicted the existence of a 'weak neutral current' in
the nucleus of an atom. In 1973–4, weak neutral currents were actually detected at two particle accelerators in Europe and the USA; the
Weinberg theory became generally accepted; and in 1979 Weinberg

shared the Nobel prize. The combined theory of all the forces except gravitation is now known as 'the standard model' of matter.

A greater prize, however, is still up for grabs—the understanding of what happened in the early moments of the universe, when gravitation was unified with the other three forces. According to the theory of the Big Bang this unification happened under conditions of unimaginably high temperature and energy, that make those inside the atom bomb pale into insignificance. To go from understanding atomic energy to explaining the Big Bang, Weinberg says, 'we now have a larger energy gap to leap over than physics had in going from Democritus to the discovery of atoms at the beginning of the twentieth century: and it may take us that long—2,500 years—but I don't think so.'

The hope remains string theory. String theory replaces the idea of the elementary particle as a point with the idea of it as a line or loop, a 'string'. The states of a particle may be produced by sending waves along this string, much like the overtones produced by a vibrating violin string. But cosmic strings keep vibrating forever, since they are not composed of atoms or anything else, only space itself.

String theory is formidably difficult—involving ten or more dimensions (as opposed to Einstein's four)—and progress has been disappointing. 'There's now a theory which is supposed to be deeper than string theory which manifests itself as string theory on various approximations, called M (for membrane or matrix) theory,' says Weinberg. 'It's very fragmentary. There isn't a set of principles. We have a set of approximations to a theory without having the theory.' But, he maintains hopefully, 'great science is done that way. It's very rare that you have something like general relativity, where some great genius, like Einstein, formulates a fundamental principle, then works out the mathematical consequences and publishes them in a series of articles.'

It irritates Weinberg to be reminded of a historian of science who recently called string theory 'the best-ever form of Platonic mysticism'.

'That's crap,' he says forcefully. 'The string theorists are trying to develop precise mathematical theories whose predictions agree with experiments. There's nothing mystical about it at all, if mystical means transcending the ordinary processes of reason. It's real cutting-edge stuff. I wish I had invented it.'

His irritation is a reminder that Weinberg has had his flashpoints, not only with theologians, but also with sociologists and philosophers of science. He has been particularly scornful of speculations about science as just one more form of knowledge, no more or less 'true' than other ways of understanding nature. Science, say some theorists, is a belief system and the truth or falsity of scientific statements is determined by the culture in which the scientists making those statements work. Philosopher Paul Feyerabend even went so far as to call science a 'superstition'.

Although Weinberg studied philosophy devotedly in his early college years and is widely read in the classics, he devotes a whole chapter of *Dreams of a Final Theory* to an attack on philosophy in science. In the book he writes: 'I know of *no one* who has participated actively in the advance of physics in the postwar period whose research has been significantly helped by the work of philosophers.'

Nor does he accept that there are strong links between cultures and the physics they practise. He is dubious that the disorder of German society after the World War I had much influence on the development of quantum mechanics by Heisenberg and others, as suggested by historian of science Paul Forman. 'To me the logical progression into quantum mechanics is so compelling. I mean quantum mechanics was there to be discovered and it was sucking people into it; I tend to see these scientific theories as pre-existing whirlpools which draw us in.'

That is why Weinberg is unsympathetic to feminist views of science. He wants to see more women physicists than in the past— 'Today there are several women whose grasp of the mathematics of string theory I find awesome; I couldn't possibly match it'—but he does not believe that their being women can alter the nature of physics. 'I think that some feminists have hold of the wrong end of the stick. They see the opening of the hard sciences to women as something that's going to change the sciences. If that were true, it would provide a rational basis for keeping women out.' He compares the situation for women in science today with that for the Jews. 'All the great men of science before 1800 were not Jewish. But no one would ask the question today, "why have so few Jews contributed to science?" Maybe 100 years from now no one will think of asking that question about women.'

Hardly surprisingly, he was delighted by the now-famous hoax article published in the journal *Social Text* by the physicist Alan Sokal in 1996. Sokal's article was a parody of relativist philosophy. It cobbled together quotes from influential French thinkers such as Jacques Derrida and concluded that the theory of quantum gravity could be used to support a left-wing political line. Although the article was nothing more than pretentious nonsense, the journal's editors accepted and published it in full.

Weinberg wrote an analysis of the issues Sokal's article raised in *The New York Review of Books* and then responded to the letters that followed, concluding with the remark that he had wrestled with what the French philosopher Jacques Derrida might have meant by 'the Einsteinian constant' only to decide that 'Derrida in context is even worse than Derrida out of context'. One mathematician colleague, whose education was in the continental tradition that mixes philosophy with science, felt that he had not been fair to Derrida, says Weinberg; otherwise, there was a very positive reaction, especially from working scientists.

But does he think he persuaded any of the postmodernists and admirers of Derrida? 'No, I don't think so. I long ago gave up the idea that I was going to persuade anyone. It seems to be part of a whole way of life, you know. "Every boy and every gal, | That's born into the world alive, | Is either a little Liberal | Or else a little Conservative",'

Weinberg quotes from Gilbert and Sullivan. 'There are people who are outside the argument, however, who look in; they may be persuaded on one side or the other.'

by Andrew Robinson

A Theory of Everything

I would hope that in the next century there will be an answer to the question 'What are the Laws of Nature?' That is, what are the deepest physical principles, from which all other universal scientific truths may potentially be deduced?

It's an easy question to ask, but not so easy to answer. The final physical laws may not be known for several centuries to come, if ever. There is, however, a problem of narrower scope, which I think will be solved in the first decades of the twenty-first century, and that will take us an important step towards a final theory. It is to learn the origin of the mass of the known elementary particles: the electron, quarks, and so on.

Our best theories suggest that these masses arise from a field that pervades the universe, and whose strength settled into its present value when the Big Bang was less than a second old. Experiments at the next generation of large particle accelerators are likely either to verify this idea, or one of the alternatives that have been suggested.

When we learn the details of the mechanism that gives mass to the elementary particles we may come to understand why these masses are so diverse: why the electron, for instance, weighs 350,000 times less than the heaviest quark. Even more important, we may begin to learn why all these masses are vastly less than the Planck mass, the one quantity with the units of mass that can be formed from the fundamental constants of quantum mechanics and general relativity.

None of this is likely to have any direct applications to technology. The largest impact of such advances in fundamental physics will be cultural: they will reinforce the view that nature is governed by impersonal laws, laws that do not give any special status to life, and yet laws that humans are able to discover and understand.

Further Reading

Weinberg, Steven, *The First Three Minutes: A Modern View of the Origin of the Universe* (New York, 1977).

—— *The Discovery of Subatomic Particles* (San Francisco, 1982).

—— *Elementary Particles and the Laws of Physics: The 1986 Dirac Memorial Lectures*, with R. P. Feynman (Cambridge, 1987).

—— *Dreams of a Final Theory: The Search for the Fundamental Laws of Nature* (London, 1993).

Slavoj Žižek

'SHIT CAN SERVE AS A *MATIÈRE-A-PENSER*,' WRITES SLAVOJ ŽIŽEK IN *THE PLAGUE OF FANTASIES*. 'IN A TRADITIONAL GERMAN LAVATORY, THE HOLE IN WHICH SHIT DISAPPEARS AFTER WE FLUSH WATER IS WAY IN FRONT, SO THAT THE SHIT IS FIRST LAID OUT FOR US TO SNIFF AT AND INSPECT FOR TRACES OF SOME ILLNESS; IN THE TYPICAL FRENCH LAVATORY, ON THE CONTRARY, THE HOLE IS IN THE BACK, WHILE THE ANGLO-SAXON LAVATORY PRESENTS A KIND OF SYNTHESIS—THE BASIN IS FULL OF WATER, SO THAT THE SHIT FLOATS IN IT— VISIBLE BUT NOT TO BE INSPECTED.'

These national variations in plumbing reveal the way in which even the most basic human functions—dealing with excremental excess—are pervaded by psychological trauma and ideology. 'The moment an academic visits the restroom after a heated discussion he is again knee-deep in ideology.'

It is writing like this that has made Slavoj Žižek, (pronounced Slavoy Sheeshcck), a philosopher from Slovenia, a cult figure across the United States. Žižek can analyse the latest Hollywood block-buster, Immanuel Kant, the Bosnian crisis, and the Freudian death-drive in the same breath, an intellectual bricolage that endears him to the Tarantino generation. He is also shockingly politically incorrect—his book *The Metastases of Enjoyment*, which flaunts a cover picture of a female anatomical figure cut open to reveal her organs, was recently banned by the feminist bookshop Silver Moon.

Žižek's books (twelve in the last nine years) are swapped like samizdat texts among trendy academics while his public appearances attract flocks of devoted followers. When he spoke at Harvard University in 1998, the audience was so large that the campus police had to be summoned to control hysterical fans being turned away at the door.

But beneath all the razzmatazz and the international star status, has Žižek got something serious to say? Has he become the exotic toy of American universities? Or is his one of the most important voices emerging from the collapse of the Soviet experiment, the release of countries like the former Yugoslavia from the Communist yoke?

We met, not in the artificial glare of the international lecture circuit but in Žižek's flat, in a modern apartment block built since Slovenian independence in his home town of Ljubljana. Between visiting fellowships in America, he stills holds a position as senior researcher at the town's Institute for Social Sciences.

His work, Žižek explains, covers three main areas. He studies German idealist philosophy—Kant, Hegel, and others; he analyses political mechanisms and how power functions; and he dissects popular culture, films in particular. His thinking is shaped by his lifelong commitment to the philosophy of Jacques Lacan, the notoriously complex French psychoanalyst who combined the legacy of Sigmund Freud with structuralist ideas about language in order to explain the paradoxical workings of the unconscious.

Žižek uses popular culture to explain Lacan's complicated ideas. But he also uses Lacan to critique popular culture. One of his most popular books is an edited collection of essays on Hitchcock's films entitled *Everything You Always Wanted to Know about Lacan (But were Afraid to Ask Hitchcock)*.

How do all these apparently divergent areas of interest relate, I ask Žižek. Is it in the Lacanian notion of 'impossibility'? Yes, he agrees, and goes on to explain what Lacan meant by impossibility. Lacan believed that we all live in the realm of language and symbolic representation and that even the unconscious is 'structured like a language'. Just as language always tries to represent reality but can never quite capture it adequately, so we can never know our real inner feelings and fantasies but instead fabricate them in a series of images or fetishistic substitutions. Žižek explains: 'Because we are symbolic animals living within the domain of language, we cannot say it all. There is a certain failure of language and it's there that you touch the Real.' This does not mean that Lacan's Real, or Impossible, is a mythical notion, located in some other-worldly site. Rather, it is grounded in politics and social mechanisms.

Žižek, typically, explains this further by talking about the film *Titanic.* 'How is the catastrophe connected to the couple, the rich upper-class girl and the poor lower-class boy? At what exact moment does the iceberg hit the ship? After making love, they go up on the deck and embrace again and then she tells him "I will stay with you and abandon my people". At that moment the iceberg hits the ship. What's the point? I claim the true catastrophe would have been for them really to stay together, because it wouldn't work and they would split. It's in order to save that impossible dream that the ship must sink. The impossibility, to put it in Lacanian terms, is the impossibility of the sexual relationship. It is to conceal that impossibility that you must have this big tragedy.'

According to Žižek, the lessons of Lacanian psychoanalysis can help us understand contemporary culture. Because we cannot get in touch with our real feelings, according to Lacan, we represent them by symbolic images and kid ourselves that these are really our fantasies, while secretly feeling relieved that some sort of control over our turbulent emotions has been established. 'The prohibition of desire in order to be operative must be eroticized. The regulation of desire leads to the desire for regulation itself,' says Žižek.

This paradox is central to Žižek's interpretation of contemporary culture. While we inhabit what is apparently a hugely permissive society, which allows sexual freedom, grants rights to ethnic minorities, and upholds free speech, we actually live within a regime of self-imposed, hidden, and eroticized prohibitions. 'Precisely by dwelling in this postmodern world, effectively your life is much more regulated,' he says. In the old days one was compelled to do things by threats and legislation. Now 'beneath the appearance of free choice, it's a much more severe order because it terrorizes you from within.'

Contemporary politics is similarly constrained. Tony Blair and Bill Clinton make minor changes to the style and presentation of public life while failing to tackle broader questions of how society should be governed. Žižek thinks that genuine political action is virtually impossible now because capitalism has won the ideological war and nobody is seriously questioning its values or rules. But he argues that, just as postmodern culture persuades itself that it lives in an age of freedom, so politicians mask their real limitations with a façade of energetic political activity.

'In the old days of essentialism (the naive belief that it was possible to impose one universal set of political values), the left focused too much on simple economic issues and the primacy of the class struggle,' Žižek argues. 'Now the days of essentialism are over and instead of the one struggle, you have plurality, gay rights, ecology, ethnic identity, whatever. I nonetheless claim that the price paid for this apparent plurality is that something has been excluded. Nobody on the left really thinks about a global alternative to capitalism.'

Žižek devotes much of his work to uncovering the prohibitions and constraints of contemporary politics and culture. He argues that if we understand the paradoxical covert nature of pleasure and regulation in our society, we may become emancipated or at least less naively vulnerable.

But beyond this task of consciousness raising, Žižek also thinks that an authentic political act is still possible. It is a stance that sets him apart from other international theoretical thinkers, who are more pessimistic about the possibility of breaking out of the system of rules that govern political action. 'An authentic political act is never just an act within a set of rules. It's an act which retroactively changes and establishes the rules of its own possibility. You do something which appears crazy and impossible. But your very intervention changes the rules themselves.' Former US president Richard Nixon's visit to China is the example he cites.

'My message is not a hopeless one,' stresses Žižek. 'Unlike the academics of the Frankfurt School I do not argue that although we seem to be free we really operate within a system that is totally rule-bound. No, no. My argument is that psychoanalysis is optimistic. It teaches us that if powerful institutions are to work, the exercise of their power must be inconsistent. And it is the gap between what these powerful institutions state publicly and those prohibitions which are implicit but never voiced which opens up a space for change.'

Žižek's 'gap' can best be understood in terms of the paradoxical Lacanian notion that real prohibitions and desires are camouflaged and expressed by licensed fantasies. In the days of Stalin, he explains, not only was it forbidden to criticize Stalin, but it was also forbidden to publically admit that you were not allowed to criticize Stalin. The real rules were hidden by apparent freedom and by the expectation that you would not want to break them anyway. Power works in these strange and insidious ways, he argues, and so sometimes it can be overturned.

According to Elizabeth Bronfen, another psychoanalytic theorist, it is Žižek's acute political analysis, reading 'political events as part of the image repertoire of popular culture' and revealing 'patterns of thinking' which span high politics and low culture, which makes his work so important. Unlike celebrity American theorists like Judith Butler, Žižek is grounded in the urgent demands of practical politics. Unlike them, he is prepared to envisage the possibility of revolutionary activity but he is also much more sensitive to its devastating effects.

Žižek's awareness of the demands of revolutionary activity was shaped by recent Slovenian history. As we wander off for lunch he shows me one of the streets near his flat where there was some shooting during the Slovenian breakaway from the Yugoslavian federation. In fact the ten-day war with Yugoslavia in 1991 was rather a non-event, as far as he was concerned. His main memory of it was that he was completing the Hitchcock book, and found difficulty swapping proofs with his other Slovenian collaborators because of the city blockades. His other memory was that Ljubljana main television station showed the American drama *Twin Peaks* every night for a week, a way, he thinks, of taking the public's mind off the situation.

In fact Žižek is quite keen to play down his Slovenian background. His British publishers Verso wanted to package one of his books as 'written from the heart of the Yugoslavian conflict', but Žižek refused. According to Miran Božovič, another of the Slovenian Lacanians in the group Žižek leads, a number of Yugoslav academics have taken advantage of their country's situation to grab attention for themselves. This flaunting of ethnicity annoys the Lacanian group. Indeed they are not enamoured of Slovenian culture at all—its poets and filmmakers are not interesting, they say—but are hungry for Western ideas, British novels, American films.

And yet the political urgency of Žižek and his Lacanian group is directly related to Slovenian politics. They were initially attracted to Lacan—Žižek went to Paris in the early 1980s where he worked with Lacan and Jacques-Alain Miller, Lacan's chief sidekick—precisely because his thinking worried the establishment. Communist politicians who relied upon managing the economy and knowing everything about people's lives were naturally troubled by the main tenet of Lacanian psychoanalysis—that the unconscious is unknowable, that it represents the empty place of power. And so in the last years of the communist regime, Lacanian philosophy became directly connected with dissident politics and alternative subculture.

The Slovenian Lacanians became the official ideologists for the Punk movement, writing lyrics for them, and organized a visit to Ljubljana of Jacques-Alain Miller, which pulled in record crowds. It was, Žižek thinks, a Golden Age, a heroic period which he looks back to with some nostalgia.

Even now, although, post-independence, Slovenian politics has quietened down and the Lacanians are no longer directly involved in public debate, Žižek is still himself involved in politics. He helped to found the Liberal Democratic Party, now the ruling party, and in 1990 he ran for the presidency and was narrowly defeated. He made a decision later to concentrate on theory rather than fulltime politics, but continues to attend party conferences and writes speeches. Later, over dinner at the Restaurant Marilyn Monroe, beneath the cinema where he watched Hollywood films as a child, he reveals more. He has the ear of the prime minister and was approached to run the Slovenian equivalent of MI5. People, he says, think it odd that as a philosopher he should be interested in the realpolitik of public life but in fact he loves the dirty backroom deals which are the stuff of politics.

Slavoj Žižek is an enigma. In many ways he plays on just the image of the East European, eccentric philosopher figure which he claims to disown. During his visiting professorships in America he goes to great lengths to avoid seeing students, inventing names and appointments so that it appears that his office hours are all booked up. An American, he says, would not get away with it, but he can happily be idiosyncratic.

Similarly, he tells me enthusiastically of his various ploys in Ljubljana for avoiding work. He has always encouraged his Ph.D. students to write their own appraisals, merely signing his name at the bottom of their account. But now he has even started getting the students to forge his signature, and the one time recently that he actually did put pen to paper, the authorities contacted him because they thought his real signature looked like a forgery. In fact the amount of administrative or teaching work that Žižek has to do is minimal anyway. His post at the Institute of Social Sciences, created during the communist regime and designed back then so that he would have no subversive contact with students, only requires him to produce research publications. He rents a small studio office and puts in an appearance at the Institute about once a month. 'I am really on permanent sabbatical,' he says.

But for all the eccentric philosopher tales—and they are highly entertaining—there is a serious side to Žižek. He possesses a passionate seriousness and intellectual power which compels Terry Eagleton, Warton professor of English literature at the University of Oxford and Britain's chief ideology theorist, to call him 'the most important cultural and psychoanalytic theorist now writing.' Elizabeth Bronfen describes him as 'leading the next wave' in international thinking, now that Foucault and others have died. He also has a determined independence, which is waking up to the way in which he is being packaged by publishers and is starting to resist. He refuses to move fulltime to the States, preferring to stay in Ljubljana where he is loyal to his Lacanian group of friends. And he is putting a stop to the stream of popular books with outrageous jokes and lurid covers. His book, *The Ticklish Subject*, published in 1999, contains 400 pages of dense philosophical analysis. The cover is just a simple white feather on a black background. Not even Silver Moon could object to that.

by Jennifer Wallace

Closing the Gap

SLAVOJ ŽIŽEK

n Andrew Niccol's futuristic film thriller *Gattacca*, Ethan Hawke and Uma Thurman prove their love by throwing away the lock of hair they have given each other. In this society of the future couples exchange hair samples so that a potential partner's genetic quality can be established by scientific analysis. Access to a privileged social class or elite is established 'objectively', through genetic analysis of the newborn.

For me, *Gattacca* merely extrapolates the prospect which can already be glimpsed today—of legitimizing social authority and power in terms of the genetic code. By eliminating artificial forms of inequality, founded on tradition and culture, we look set to open up the way for a new hierarchical order based on the genetic make-up of individuals.

But might there be a way out of this destiny? Can we arrange for people in the future to be what they want rather than what their genes dictate? What if we adopted a strategy of deliberately refusing to know too much about our biology?

At present we live in a period of tension. On the one hand, there are those who explain the origin of humanity in narrow scientific terms; we evolved via natural selection, our brains are the sum of cognitive mechanisms, etc. On the other hand, a multitude of spiritualisms (from traditional religion to New Age cosmology) refuse to accept that man can be reduced to 'mechanic' nature.

This spiritual fear of reducing people to scientific explanations (which is evident in the public's widespread resistance to human cloning) is based on a suspicion that a complete explanation of man in terms of biology would deprive human beings of their basic dignity and freedom.

So the breakthrough I would most like to see happen in the

twenty-first century is the unification of the natural sciences with the notion of human freedom. Perhaps it will be possible to elaborate how the radical contingency discernible in prehuman nature itself (it is clear from quantum physics, for instance, that there are aspects of the physical nature of the world that we just don't understand) is linked to that gap or 'void' in human nature that allows for human freedom.

This gap is explained by Lacanian psychoanalysis. According to Jacques Lacan, our subconscious desires elude our understanding and in their place we substitute licensed fantasies which appear to be wild, but which are actually regulated and regulating and comforting because they are easier to deal with than the real thing. The gap is thus between what we think we fantasize about and what we actually do fantasize about.

It is my contention that the psychoanalytic theory of human subjectivity, far from being obsolete, will play a crucial role in unifying the natural sciences and the notion of human freedom in the next century. Psychoanalysis will help because it will introduce a radical unknowability into a world which seems all too predictable as it is increasingly explained away in biological terms. Psychoanalysis is all about not being able to understand the unconscious. That unknowability spells freedom.

Further Reading

Žižek, Slavoj, *The Indivisible Remainder: An Essay on Schelling and Related Matters* (London, 1996).
—— *The Plague of Fantasies* (London, 1997).
—— *The Ticklish Subject: The Absent Center of Political Ontology* (London, 1999).

Acknowledgements

Picture Credits

Joel Chant: pages 212, 226; Jacky Chapman; 258, 290; Ian Cook: 2; Jez Coulson: 22; Dan Dry: 200; Geoff Franklin: 86, 106, 168, 180, 280; Steven Klein: 268; Robin Mayes: 300; Paula McLerner: 138; Mark Pepper: 118; Michael Powell: 34, 96; Paul Schnaittacher: 158; Tone Stojko: 310; Neil Turner: 46, 66, 76, 148, 190, 236; Frank Ward: 128.

Text Credits

Introduction: © Jonathan Weiner

Chinua Achebe: profile © Jennifer Wallace; prediction © Chinua Achebe

French Anderson: profile © Tim Cornwell; prediction © French Anderson

Noam Chomsky: profile © The Times Higher Education Supplement; prediction © Noam Chomsky

Arthur C. Clarke: profile © The Times Higher Education Supplement; prediction © Arthur C. Clarke

Paul Davies: profile © The Times Higher Education Supplement; prediction © Paul Davies

Richard Dawkins: profile © The Times Higher Education Supplement; prediction © Richard Dawkins

Daniel Dennett: profile © The Times Higher Education Supplement; prediction © Daniel Dennett

Carl Djerassi: profile © The Times Higher Education Supplement; prediction © Carl Djerassi

Andrea Dworkin: profile © Jennifer Wallace; prediction © Andrea Dworkin

Umberto Eco: profile © Domenic Pacitti; prediction © Domenic Pacitti

Francis Fukuyama: profile © The Times Higher Education Supplement; prediction © Francis Fukuyama

Index

Note: **Emboldened** page numbers indicate major entries